POWER LINES

INSIDE TECHNOLOGY

edited by Wiebe E. Bijker, W. Bernard Carlson, and Trevor Pinch

A list of the series appears at the back of the book.

POWER LINES

Electricity in American Life and Letters, 1882–1952

JENNIFER L. LIEBERMAN

The MIT Press
Cambridge, Massachusetts
London, England

© 2017 Massachusetts Institute of Technology

All rights reserved. No part of this book may be reproduced in any form by any electronic or mechanical means (including photocopying, recording, or information storage and retrieval) without permission in writing from the publisher.

Set in Bembo Std by Toppan Best-set Premedia Limited. Printed and bound in the United States of America.

Library of Congress Cataloging-in-Publication Data

Names: Lieberman, Jennifer L., author.
Title: Power lines : electricity in American life and letters, 1882-1952 / Jennifer L. Lieberman.
Description: Cambridge, MA : MIT Press, [2017] | Series: Inside technology | Includes bibliographical references and index.
Identifiers: LCCN 2016052817 | ISBN 9780262036375 (hardcover : alk. paper)
Subjects: LCSH: Electrification--Social aspects--United States--History. | Electric power--Social aspects--United States--History. | Technology in literature--History. | American literature--History and criticism.
Classification: LCC HD9685.U5 L49 2017 | DDC 333.793/2097309041--dc23 LC record available at https://lccn.loc.gov/2016052817

10 9 8 7 6 5 4 3 2 1

CONTENTS

Acknowledgments vii

INTRODUCTION 1

1 MARK TWAIN AND THE TECHNOLOGICAL FALLACY 17

2 SHOCK AND SENSIBILITY: THE RHETORICS OF ELECTRIC EXECUTION 51

3 CHARLOTTE PERKINS GILMAN'S HUMAN STORAGE BATTERY AND OTHER FANTASIES OF INTERCONNECTION 91

4 THE CALL OF THE WIRES: JACK LONDON AND THE INTERPRETIVE FLEXIBILITY OF ELECTRICAL POWER 129

5 RALPH ELLISON'S AND LEWIS MUMFORD'S ELECTRIFYING HUMANISM 167

CONCLUSION 211

Notes 221
References 247
Index 265

ACKNOWLEDGMENTS

This book took shape in conversations in conference rooms, classrooms, and coffee shops. It began as a dissertation project in the Department of English at the University of Illinois, Urbana-Champaign (U of I), with the help of the intellectual community I developed there. I would like to thank, particularly, the Illinois Program for Research in the Humanities (IPRH) for the opportunity to workshop early drafts with an interdisciplinary group of peers; John Carlos Rowe, who has provided feedback on drafts since I met him at the Futures of American Studies Institute; Guy Tal, Elizabeth Hoiem, Kimberly O'Neill, Anne Brubaker, and the many dear friends who were also my sounding boards, especially Mary Unger, who has read drafts from the dissertation and the book alike; and my dissertation committee: Robert Markley, Bruce Michelson, Rayvon Fouché, and Melissa Littlefield. The lessons from my committee members remain with me. Their tutelage helped me transform my half-formed ideas into wholes and guided me to questions that I continued to pursue in my postdoctoral research and beyond.

As a postdoctoral fellow in the Department of Science and Technology Studies (STS) at Cornell University, I enjoyed the privilege of working under the mentorship of Ronald Kline. He taught me how to reframe my approach to technological history, which in turn reshaped the way I understood literary history. With perceptive feedback and engaging conversation, he helped me to evolve my dissertation into the wider-reaching research project that you see before you today. Indeed, the entire department and its superset, the Science Studies Research Group (SSRG), deserve recognition: the students and faculty who shared their ideas every week helped me to grow as a thinker and a scholar. I remain appreciative for the feedback that Sara Pritchard, Trevor Pinch, and other SSRG attendees provided when I gave talks and shared early drafts. I am similarly indebted to Joan Jacobs Brumberg, whose careful writing and engaging pedagogy offered an aspirational model in interdisciplinarity. I am thankful, also, to the Historians Are Writers! group (affectionately acronymed HAW!) at Cornell University: Aaron Sachs, Amy Kohout, Josi Ward, and others in this organization reminded me that academic arguments can be beautifully and lovingly written. In the process, they helped me to develop my narrative voice in addition to my arguments. The editors of the Inside Technology series, Wiebe E. Bijker, W. Bernard Carlson, and Trevor Pinch; my editor at the MIT Press, Katie Helke, and the manuscript editor, Paul Bethge, also helped me to realize my vision for this project in form as well as content. Thank you all.

Research for this book was made possible with funding and support awarded from the U of I; the IPRH; the Bakken Library and Museum (where I had the pleasure to work alongside librarian Elizabeth Ihrig); the Eaton Collection at the University of California, Riverside (where I was assisted in my research by Melissa Conway); the Smithsonian Institute's Dibner Library of the History of Science and Technology (where Head of Special Collections Lilla Vekerdy helped me to dig up many of the fascinating texts I discuss here); the National Science Foundation (which enabled me to visit many archives, including miSci, Museum of Innovation and Science, where I was assisted by Chris Hunter); Cornell University; and the University of North Florida (UNF). Each of these institutions offered the opportunity to work with and learn from archivists and academics who have piqued my curiosity and challenged me in uncountable ways. At UNF, I received crucial feedback and support from Samuel Kimball and Brian Striar, and from my interdisciplinary writing group. Nicholas de Villiers, Stephen Gosden, Laura Heffernan, Jason Hibbard, Tru Leverette, Betsy Nies, and Sarah Provost—your insights improved every chapter of this book.

I am thankful for the suggestions from un-nameable sources, as well—including the questions and comments I received at annual meetings of the American Literature Association; the American Studies Association; the Modern Language Association; the Society for the Study of the Multi-Ethnic Literature of the United States; the Society for the History of Technology; the Society for Literature, Science, and the Arts; and the Society for the Study of American Women Writers. I am especially

Acknowledgments

grateful for the feedback from my reviewers. Although I do not know who they are, their suggestions helped me to clarify and improve my argument substantively.

At the proofing stage, I am thankful for assistance from Deborah Cantor-Adams at the MIT Press. I am also grateful for funding from the Florida Blue Center for Ethics, which enabled me to hire two incredibly talented research assistants: Kayla Hilliar and Layne Perkins. Thank you all for your help at this vital stage.

Finally, I must express my gratitude to my family for their love and support. Thank you to my parents, Barbara and Richard Lieberman, and my brother and sister-in-law, Ian and Maggie Lieberman. Thank you to my husband Colman Stiteler, for your insights, your encouragement, and your patience. I dedicate this book to you.

INTRODUCTION

I shot the current through all the fences and struck the whole host dead in their tracks! *There* was a groan you could *hear!* It voiced the death-pang of eleven thousand men.
—Mark Twain, *A Connecticut Yankee in King Arthur's Court*

That is why I fight my battle with Monopolated Light & Power. The deeper reason, I mean: It allows me to feel my vital aliveness.
—Ralph Ellison, *Invisible Man*

Some of the most arresting scenes in American letters, from the mass electrocutions that conclude Mark Twain's 1889 novel *A Connecticut Yankee in King Arthur's Court* through the electrical power siphoned to light 1,396 inefficient bulbs in Ralph Ellison's 1952 novel *Invisible Man*, arise from the meanings that characters project onto electric light and power. Both Invisible Man and the Yankee hope that an electrical age will usher in a new and rational era that replaces hierarchies of class and race with meritocracy and democracy. Both protagonists are woefully mistaken; older structures of social power are not so easily broken.

Why would Twain, Ellison, and many other prominent writers—including Charlotte Perkins Gilman, Theodore Dreiser, and Jack London—craft such thoughtful images of electricity in works that have more to do with social stagnation than with mechanical advancement? In part, their fascination was anecdotal: electricity made an impression on each of these writers' lives. But they were also drawn to electricity because it was the life force and the death spark of the day. It seemed to these writers that electricity might change everything.

Visualize this mutable world through Mark Twain's eyes. In 1884, he was traveling across the country with George Washington Cable and lecturing along the way. This speaking tour piqued the interest in Arthuriana that would later inspire *A Connecticut Yankee in King Arthur's Court*. It also offered Twain the opportunity to compare the lamp-lit Detroit of his past to the newly electrified Michigan city. On December 16 he wrote the following in his journal:

> Lots of things have changed, & always for the better. They have dry towels in the hotels, now, instead of the pulpy damp rag of the former day, which shuddered you up like a cold poultice; & they have electrical buttons, now, instead of those crooked bell handles <&> which always tore your hand & made you break a lot of the commandments. ... And then the new light—there was nothing like it when I was on the highway before. I was in Detroit last night, & for the first time saw a city where the night was as beautiful as the day; saw for the first time in place of sallow twilight bought at three dollars a thousand feet, <saw> clusters of coruscating electric suns floating in the sky without visible support, & casting a mellow radiance upon the snow covered spires & domes and roofs & far stretching thoroughfares which gave to the spectacle that daintiness & delicacy of a picture <[or] of> & reminded one of the airy <&> unreal cities caught in the glimpses of a dream. (Twain 1979, 81)

In this entry, Twain savors the comforts he can enjoy on this journey: warm towels, telephones, electric call buttons. When he discusses the city's electric lighting, he grows rhapsodic.

Twain's change in tone corresponds to a change in the scale of the system he describes. He recounts the small circuit he personally controls with a humorous, realistic comparison. He characterizes the electric button as a perceptible improvement by exaggerating previous discomforts ("crooked bell handles <&> *which always tore* your hand"). In contrast, he portrays the large-scale lighting system as entirely new, rather than as an improvement upon an earlier form of illumination ("there was nothing like it when I was on the highway before"). The Detroit lighting system that astonished Twain was indeed remarkable; 150-foot-high arc-light towers could be considered sublime even by today's standards (Jakle 2001, 48–50). Yet Twain doesn't reflect upon the impressive infrastructure of this tower system. In fact, he erases its materiality by imagining the electric light "floating in the sky without visible support." To express the sense of progress this scene evokes for him, he packages his empirical observations in the fantastical language of an "unreal" dream.

Twain's interpretation was misleading but understandably attractive. The historians of technology David Edgerton (2006), David Nye (1992), and Wolfgang Schivelbush (1988) have shown that electrification changed American cityscapes more incrementally and inconsistently than Twain allows.[1] Still, despite the glaring imperfections of actual electrical development, fantasies like this one proliferated. The idea that electrification created exclusively modern experiences was enticing because it corroborated the prevalent impression that the turn of the twentieth century was a moment of radical upheaval—a sense many Americans shared, even if they hotly debated whether the changes they perceived were, as Twain claimed in this journal entry, "always for the better." Whether Americans put their faith in electricity or feared its adoption, it became a metonym for American life at the end of the nineteenth century.

ELECTRICAL POWER AS A DISTINCTIVE SYMBOL

Before electricity became a supple symbol of American modernity, trains and telegraphy had already promised to sever the modern present from the recent past by "annihilating" space and time (MacDougall 2013, 9). Those systems often employed electricity as a power source, and they became intimately associated with it. In fact, three recent studies have examined electrical communication as a theme in nineteenth-century and twentieth-century American literature: Sam Halliday's *Science and Technology in the Age of Hawthorne, Melville, Twain, and James: Thinking and Writing Electrically* (2007), Paul Gilmore's *Aesthetic Materialism: Electricity and American Romanticism* (2009), and Mark Goble's *Beautiful Circuits: Modernism and the Mediated Life* (2010). These books offer innovative arguments about how electric communication systems inspired new paradigms of communication that evolved in tandem with the conventions of American literature.

As Gilmore, Goble, and Halliday also demonstrate, the cultural meanings of electricity and electrical communication systems were interrelated. At the turn of the twentieth century, Americans imagined that those systems signified progress or, conversely, the frenetic pace of life in an increasingly corporate economy. While some supposed that "electric nerves" (including transportation, communication, and power systems) could connect humankind into one large social body, others feared that the same systems would fragment traditional domestic relationships. Despite their overlapping connotations, these systems were not interchangeable. Communication systems functioned differently than electrical power systems both in the literary imagination and in the world. The low-voltage telegraph and telephone inspired fantasies about altering space and time, while high-voltage electrical systems provoked comparatively grandiose dreams about controlling Nature and life itself (in the respective forms of electrons and animal electricity).

Consequently, although some communication and transportation systems employed electricity, electrical power signified something different and exceptional. A force that could power anything from a submarine to a vacuum cleaner, electricity encouraged the fantasy that inventors could create any device that could be imagined.[2] Literary and technical writers anticipated that electrical systems could accomplish anything from eliminating hunger to instantaneously extinguishing life. Electricity seemed so powerful to Americans at the time because it was uniquely multifarious. Invisibly permeating the atmosphere but also harnessable for button-operated use, electricity bridged the perceived divide between the human-built environment and the natural world.[3]

The Janus-faced quality of electricity rendered it a thrilling energy to use and to tinker with, as historians have demonstrated. This ambivalence also made electricity a fascinating conceptual tool that inflected how many Americans described and understood their world. As Paul Edwards notes, "All tools, including clocks and computers, have both practical and metaphorical or symbolic dimensions" that evolve in feedback with one another, because "the experience of using any tool changes the user's awareness of the structure of reality and alters his or her sense of the human possibilities within it" (1997, 29). The metaphorical dimension of electricity was irreducibly plural and thus was particularly captivating. If tools "symbolize the activities they enable" (quoted in ibid., 29), then electricity signified multiple, contradictory ideas at once—spiritualism[4] and utilitarianism, death and re-invigoration, and human control and unwieldiness, to name only a few.

In this book, I explore how incompatible conceptions of electricity paradoxically coexisted in the American cultural imagination. I also analyze the stakes of these conceptualizations. Electro-medical treatments and the electric chair emerged from such disparate perceptions of electricity, as did the enduring poetry of Walt Whitman's "body electric" and of Bessie Smith's sultry blues number "Send Me to the 'Lectric Chair." These divergent applications and representations of electricity raise a question that can be answered only with a creative, interdisciplinary approach: How did such an indeterminate, contested force become a common metonym for modernity? Or, put another way, why did writers invoke electrification, with all of its ambiguity, to characterize nontechnical aspects of their times?

The association between electricity and modernity was not as natural or obvious as it might seem from the perspective of our electronics-obsessed age. This correlation evolved over decades, as *fin de siècle* writers revised the spiritualistic symbolism they inherited from Enlightenment, Romantic, and Transcendentalist figures. In the eighteenth century and in the early years of the nineteenth, electricity was most often aestheticized as a pervasive force or as charged ether. Late-nineteenth-century and early-twentieth-century writers continued to conjure these images, but they also layered into their narratives new depictions of electricity as a symbol of industry or complexity. In some cases, as in Twain's journal entry, these later writers found the conventions of fantasy sufficient to describe their impression of an electrical invention. More often they wove these romantic depictions together with other narrative modes to convey the plural and sometimes conflicting meanings they read onto electricity itself and the apparatuses it animated. Ultimately, electricity was distinctive because it was polysemous. Writers seized upon this emblem because it already had a rich aesthetic legacy *and* because its new industrial applications correlated this energy with interconnection and action at a distance. The pastoral and romantic strategies for depicting electricity as a universal force were not displaced.[5] They were built upon.

THE RISE OF SYSTEMS THINKING AND THE RESILIENCE OF INDIVIDUALISM

In this book I examine electrical power as a distinctive concept in American literary, cultural, and technological history, with connotations that have been obscured by its association with communication systems.

Nonetheless, I attribute the changing social meanings of electricity, in part, to a broader cultural fascination with systems that gained purchase at the turn of the twentieth century.[6] Scientists had investigated electrical and magnetic phenomena for generations, but electrical energy first started to enter American businesses, homes, and literature at a moment when the problem-solving strategy we might now call "systems thinking" was becoming prominent in American culture.[7] The correlation between electricity and the abstract concept of the system helps to explain why this energy became a common metaphor or metonym in American letters.

According to the historian of technology Thomas Parke Hughes (1979), Thomas Alva Edison's most significant contribution to the history of electricity was his ability to frame illumination as a systemic problem that involved multiple components (technical, as well as economic and social). Hughes studied systems in order to develop a robust social theory of technology that could account for the creation and popularization of large-scale inventions. But, as Hughes noted later in his discussion of Frederick Winslow Taylor and other figures of the Progressive Era, this tendency to see the world in systemic terms was not limited to the builders of actual technologies, such as Edison (Hughes 2004, 184–248).

In fact, turn-of-the-century writers helped to forge the language we colloquially use to describe networks today. The verbs *interconnect* and *interrelate* were coined in 1865 and 1888, respectively, and the word *network* garnered new connotations at the turn of the twentieth century.[8] The word *system* remained largely unchanged during that period. It connoted collections of objects, people, or principles—as it had for several hundred years. But new and more specific usages emerged in the mid nineteenth century and the early twentieth century. The phrase "out of one's system" emerged in 1866, "system-wide" appeared in 1902, and "system operator" was first recorded in 1903. The emergence of such terms and the proliferation of documents that use them indicate the rise of a dominant ontology that emphasized contingency and complexity over a simplistic view of machine-like causation.

The reverberations of this trend can be felt across myriad disciplines beyond the fields of literature and electrical development on which I focus. The field of ecology, which studies the relationships among various organisms and environments, was formalized during the 1860s.[9]

Meanwhile, sociology and anthropology began to focus upon complex webs of human relations. In some cases, this move toward system thinking was directly inspired by electrical science, as when proponents of electro-vitalism drew analogies to electrodynamics to understand the complex essence of life.[10] In other cases, systems thinking was influential without any direct reference to electrical science. In mathematics, for example, the failed move toward complete axiomatization seemed to suggest that our world could not be described as a sum of coherent, discrete phenomena.[11] At the turn of the twentieth century, intellectuals and public figures re-imagined their universe in terms of multivariable interactions; non-causality and interdependence became *a priori* assumptions in many disciplines; interconnection became a prominent trope in American letters; and electricity became one of the many concepts to be affiliated with, and understood through, these terms.

Historians of technology have recognized the increasing emphasis on interconnection and its association with electricity in American letters at the turn of the twentieth century. For example, Nye (1992, 19) postulated that "perhaps electrical metaphors emphasized integration and connection because electrical lines created a permanent link between a producer and a consumer, and this line opened up so many new possibilities." However, with a few notable exceptions, including Goble, Gilmore, and Halliday, American literary historians have neglected to theorize the turn-of-the-century fascination with systems that has attracted the attention of scholars in other disciplines. Instead, for decades, the theme of individualism has largely circumscribed the study of turn-of-the-century American literature.

As the literary historian Nancy Glazener argues (1997, 18), "When public school students are taught ... that the fundamental categories for fictional plots are 'man vs. man,' 'man vs. society,' 'man vs. nature,' and 'man vs. himself,' the perpetuation of individualist readings habits—not to mention gender bias—makes it difficult for any text to introduce readers to collective registers of experience and agency bigger than any individual but smaller and more specific than 'society.'" This tendency to emphasize individualistic conflicts has unduly constrained the study of American literature. It has enabled teachers and scholars to neglect the writings by major authors that deal explicitly with the theme of interconnection. As a result, students of English know Charlotte Perkins Gilman for her haunting depiction of neurasthenic isolation in her first short story, "The Yellow Wall-Paper" (published in 1892), but they are less likely to know that she also wrote prolifically about networked relationships. They remember Jack London for his short pieces about virile men confronting austere nature, while they routinely overlook his novels about the electrification of the West. Literary histories that disregard the latter texts selectively accentuate the place of the alienated individual in American letters, constructing a false cultural memory.

This issue is not relegated to English classrooms. Other disciplines have similarly accentuated the rise of individualism. According to the social systems theorist Niklas Luhmann, sociological theory shares this bias: it "conceives history as a process of increasing individualism" (1990, 107–108). Luhmann argues that theorists have considered interconnectivity only insofar as it helps them understand the individual, citing Émile Durkheim, Georg Simmel, and George Herbert Mead's formulations of the individual as a "collage" or as an "emerging unit—emerging not from history but from social encounters" (ibid., 107–108).

The overemphasis on the individual limits the questions scholars ask about life and literature; it alienates readers who are interested in the complex dynamics of systems; and it fails to account for many provocative texts, themes, and techniques that could expand our understanding of the interplay among literature, culture, and technology. In this book, I address these issues by recovering literature that employs electricity as a metaphor for more diffusive systemic themes. Still, given the prominence of individualism in our philosophical and literary traditions, moving beyond the academy's longstanding fascination with individualism entails more than applying up-to-date systems theories to our readings—and it involves more than adding texts about interconnection to the ever-expanding literary canon. This process requires an overhaul of inherited values. Many of the texts I discuss employ nontraditional rhetorical strategies in their attempts to describe systemic interconnection. Some have been forgotten because they fail to meet the criterion of aesthetic coherence upon which the study of literature was founded.[12]

Consider, for example, the systems-obsessed writings of Charlotte Perkins Gilman, which I will address more extensively in chapter 3. She draws upon the newly interconnective connotations of electricity in her nonfiction volume *Human Work* in order to promote a socialistic worldview. She builds to this point gradually, quoting a passage from Walt Whitman's "Song of Myself" (1855) to demonstrate that "current literature is full of this social reversion to-day, this 'call of the wild,' this tempting invitation to give it all up and go back to the beginning" (Gilman 1904, 103).[13]

Gilman claims to empathize with Whitman's (and implicitly London's) individualistic desire to return to nature, but she celebrates interconnectedness as a more fulfilling endeavor. She characterizes this alternative with allusions to electricity as a force: "The young human creature, as he begins to grow from the individual animal into social life, feels this intense current of force, the vast and varied desires, the vaster energies" (1904, 103). Such appeals to electricity become increasingly explicit throughout *Human Work*. Gilman concludes her study by extolling interconnection as a physical, electrical sensation: "To feel the extending light of common consciousness as Society comes alive!—the tingling 'I' that reaches wider and wider in every age, that is sweeping through the world to-day like an electric current. ...To feel the power! The endless power!" (389). Both of the passages that discuss electricity omit conjunctions, cultivating a sense of mounting excitement for "energy" and "power" without clarifying how these "forces" emerge from or represent social interconnectedness.

By beginning her argument with a critique of Whitman and "current literature," Gilman hints that a new literary movement could help Americans outgrow their obsession with individualism. Although she never specifies how this new literary form would look, she articulates her investment in interconnectivity by weaving generic conventions together and by mixing metaphors. Her descriptions of nerves and energies, of sinuous currents and vibrating power, stand in for the alternative to individualistic literature that she seeks. She seems to promote a narrative style that considers the perspectives of individuals, systems, and society in a manner akin to the multiscalar approach that Edwards (2003) advocates in his study of infrastructures.[14] Ultimately, Gilman urges writers to try to tell the story of human interconnectedness even if it can only be half-told, approximated through pieced-together analogies, blended perspectives, and metaphors drawn from electricity.

Gilman was not alone in her celebration of interconnected, electrical forces as an inspiration for artistic and social change, nor was she alone in the rhetorical strategies she used to represent social, biological, and technical systems. Many of her contemporaries drew similarly suggestive analogies to describe changes in modern life that seemed to defy first-person narration. While a number of the texts that address these issues have been under-examined by recent scholars, one related example stands out for the attention it has received: Henry Adams's 1900 essay "The Dynamo and the Virgin." In that oft-quoted piece, Adams describes himself in the third person as he ponders how the invention of the dynamo affects his individual life and his conception of historical progress:

> Satisfied that the *sequence* of men led to nothing and that the *sequence* of their society could lead no further, while the mere *sequence* of time was artificial, and the *sequence* of thought was chaos, he turned at last to the *sequence* of force; and thus it happened that, after ten years' pursuit, he found himself lying in the Gallery of Machines at the Great Exposition of 1900, his historical neck broken by the sudden irruption of forces totally new. (Adams 1918, emphasis added)

Adams's distant narrative voice and his repetition of the word *sequence* underscore the failure of the individual's apprehension of time. Adams concludes this periodic sentence with an allusion to hanging ("his historical neck broken")—a provocative analogy in view of the role that electricity played in replacing hanging with a new form of execution.

In the passage quoted above, Adams stages a violent contest between his embodied understanding of history (denoted by the mixed metaphor "historical neck") and electricity (which serves as a metonym for all "forces totally new," including radium, x rays, and other non-electrical energies). In so doing, he dramatizes the failure of the historian's first-person perspective within a new regime of "irruptive force." As the literary scholar Jennifer Fleissner argues (2004, 1), "Adams stands out for his decision to face the new developments [in scientific thought] squarely, to attempt an account in which natural forces and our confrontations with them receive the same weight as political events." Adams emphasizes rupture, whereas Gilman sees the opportunity for synthesis and evolution. Both believe that electricity unsettles existing conventions for understanding human life in individualistic terms.

Like Gilman, Adams contrasts Whitman's poetry with the new aesthetics that electricity represents. Specifically, Adams claims that Whitman's unconventionality and raw emotion cannot account for the complex meaningfulness of the dynamo:

> Adams began to ponder, asking himself whether he knew of any American artist who had ever insisted on the power of sex, as every classic had always done; but he could think only of Walt Whitman; Bret Harte, as far as the magazines would let him venture; and one or two painters, for the flesh-tones. All the rest had used sex for sentiment, never for force; to them, Eve was a tender flower, and Herodias an unfeminine horror. American art, like the American language and American education, was as far as possible sexless. Society regarded this victory over sex as its greatest triumph, and the historian readily admitted it, since the moral issue, for the moment, did not concern one who was studying the relations of unmoral force. He cared nothing for the sex of the dynamo until he could measure its energy.

Previous scholars have examined the anti-modernism that pervades such passages by focusing on what Adams says.[15] Here I am more interested in how he says it. In the excerpt quoted above, Adams constructs a compelling (if ambiguous) chimera that amalgamates "dynamo" and "Virgin," "energy" and "sex."[16] He draws electricity into a matrix of other symbols in an attempt to elicit modern themes unaccounted for within his usual modes of storytelling.[17] At the same time, he sarcastically dismisses rational rhetorical paradigms that seek only to "measure" the dynamo's "energy." Yet what alternative does he offer? What does he mean by the "sex of the dynamo"? Adams never advances a clear answer. In view of the prominence of the laws of thermodynamics in Adams's *Education*, the literary scholar Mark Seltzer argues (1992, 30) that these concepts are related metaphorically through the law of conservation: sex and electricity, like energy and matter, are forms that can be interconverted but not expunged.

While Adams appeals to these newly formulated physical laws and uses what Edwards (2003) might call a macro-scale perspective, he also alludes to an established semantic tradition. Electricity and sexuality had long been linked in American literary and cultural history, dating at least as far back as the eighteenth century, when Benjamin Franklin republished Albrecht von Haller's "venus electrificata" experiments (Franklin 1751, 10).[18] In such experiments, a young woman stands on a wax plate so as to be insulated from ground electricity. A man takes a charge from a Leyden jar and draws his lips near to hers until a static spark arcs between the two human actors. Adams conjures this rhetorical tradition but doesn't conform to it. "The Dynamo and the Virgin" doesn't depict electricity as the erotic force that attracts human bodies to one another. Instead, the dynamo inspires Adams to write history in new ways. Its "force" challenges him to think beyond his own parochial perspective—to reimagine history not as a series of human accomplishments, but as a larger drama involving systemically interconnected energies. Adams seems, in the words of Clifford Siskin (2016, 18), to sense that "retooling can produce historically different ways of knowing."

Many writers sensed, as Adams and Gilman did, that electrical power could inspire new aesthetics and new understandings of historicity. Yet the authors who strove to comprehend life in systemic terms seem to have found the task difficult to undertake and even more difficult to maintain. Even Adams didn't sustain it entirely; he doesn't alter his sequential narrative form in subsequent chapters of his *Education*.[19] Not even electricity (and the systems it came to stand for) compelled these writers to formulate a coherent literary or philosophical movement. It did, however, inspire them to play with narrative in ways that anticipate the experimental rhetorical strategies more often associated with literary modernism.[20] They dabble in bricolage, weaving together generic conventions or scales or images to evoke the impression of interconnection.

LINES OF INQUIRY

In the pages that follow, as I study these writers and the strategies that they employed to aestheticize electrical and social systems, I heed the suggestion of the novelist Saul Bellow. I urge that we set aside criticism of unclear writing and ask instead "what the 'bad writing' of a powerful novelist signifies" (Bellow 1955, 146). When Gilman or her contemporaries mix metaphors, perspectives, or genres, they are not necessarily failing to achieve coherence. More often, they are striving to envision the world as a system of human and nonhuman interconnections using rhetorical tools that are not particularly suited to the task. Their authorial techniques—and specifically their proclivity for drawing analogies to electricity—require investigation because they help revise a historical record skewed by the residual preference for works about individual struggles, because they promote ways of reading and writing that move beyond the "individualist reading habits" that Glazener identified, and because they elucidate the plural and co-constructed meanings of electricity in the American cultural imagination. *Fin de siècle* writers suggest that electricity was changing their senses of self, society, and history; they also convey the simultaneous certainty that writing could change the meanings and uses of this energy.

In the chapters that follow, I will explore the myriad ways that that Americans fashioned electricity into a metonym for modern life. Although the story of electricity is too amorphous to map onto a linear change-over-time narrative, I trace how this energy became an analogue for other social, political, and biological systems during moments that correspond to specific technical changes, including the opening of Thomas Edison's Pearl Street Station (1882), the signing of New York State's Electric Execution Act (1888), the development of long-distance power transmission systems in the American West (1890–1915), the rise of electrical consumption for leisure in the wake of World War I (1918–1925), and the federal campaign to electrify the nation (1935–1952). During each of these historical moments, electricity accrued new social meanings—often without erasing its pre-existing, already plural connotations.

I begin mapping this historical terrain with a chapter titled "Mark Twain and the Technological Fallacy," in which I argue that the introduction of the term *technology* into literary criticism of Mark Twain's *A Connecticut Yankee in King Arthur's Court* corresponds to a dramatic change in the dominant interpretation of the novel: Twain's contemporaries read *A Connecticut Yankee* as a humanist triumph, although twentieth-century scholars read it as confessional piece that conveyed the author's anxiety about technology. Noting that Twain never used the word *technology*, I interpret the rhetorical strategies he did use—especially his use of electricity as an organizing metaphor. Ultimately, this chapter contends that the anachronistic introduction of the concept *technology* into criticism of this novel reveals the paucity of critical terms we have to describe systems and technical artifacts today. I use the term *technological fallacy* to draw attention to this concern—and to urge readers to attend more closely to the idiosyncratic language historical actors used to represent and comprehend their world.

My second chapter, "Shock and Sensibility: The Rhetorics of Electric Execution," demonstrates that the technological fallacy involves more than anachronism in the interpretation of literature; I also use this concept to describe the reductive rhetoric that enables a complex social, legal, and technical process (electric execution) to seem like the result of a single material apparatus (the electric chair). This chapter draws on original archival research to argue that proponents of electric execution over-emphasized the mysteriousness and instantaneousness of the fatal electrical charge. In the process, they deflected attention from the legal and socioeconomic systems that enabled the chair to function. In contrast, prominent literary figures, including William Dean Howells, Stephen Crane, Theodore Dreiser, and Gertrude Atherton, wrote imaginative pieces that drew attention to the person in the electric chair and to the people whose actions sanctioned that person's death. By focusing on the human components of this fatal electrical circuit, these writers question why technical innovations should outweigh moral concerns. Their narratives insist that the capacity to take human lives with electrical power is not a sufficient warrant for doing so.

Atherton, Dreiser, and Howells pit vulnerable individuals against social and electrical systems. My third chapter, "Charlotte Perkins Gilman's Human Storage Battery and Other Fantasies of Interconnection," deconstructs this perceived binary by analyzing texts that emphasize the inextricability of "individuals" and "society." This chapter reads Gilman in conversation with the socialist and electrical engineer Charles Proteus Steinmetz, with the electro-medical doctor George Miller Beard, and with the sociologist Lester Frank Ward in order to examine how each of these public figures characterized electrical systems as integrative and progressive. As I noted above, Gilman criticizes American letters for its longstanding obsession with "individual versus nature" narratives. The activist-writer makes her case by drawing on the cultural currency of electrical science. She describes individuals as "human storage batteries" to assert that they are most energized when they are interlinked. In so doing, she also problematizes accounts of electrical development that omit the imperfect battery in order to portray progress as linear.

My fourth chapter, "The Call of the Wires: Jack London and the Interpretive Flexibility of Electrical Power," elaborates on the tension that Gilman identifies between the literature of individuation and that of interconnection. The tradition Gilman dubs "call of the wild literature" maps onto what literary scholars today call "literary naturalism" and what writers at the turn of the twentieth century called "new realism." Conventionally, literary scholars have characterized this influential movement by its attention to the individual who is buffeted about by "forces of nature." In chapter 4, I select London as a case study because he incorporated the theme of electrical interconnection into this famously individualistic genre. I argue that, in the process, London also revealed alternative possibilities for electrical development to his readers. By placing London in conversation with contemporary narratives by insiders in the electrical industry, this chapter modifies the canonized author's literary legacy, uncovers choices that were available to American consumers at the moment they chose to purchase and use electrical energy,[21] and undermines the industry's insistence that privately owned long-distance power transmission was the most rational and efficient way to harness electricity.

Chapter 5, "Ralph Ellison's and Lewis Mumford's Electrifying Humanism," offers a reflective case study. It argues that both Ellison and Mumford revealed the resilience of the rhetorical tradition that I delineated in chapters 1–4. Mumford imagines that electricity represents the coming of a new technological order that he calls "neotechnic." Ellison engages with Mumford's ideas explicitly and implicitly. In so doing, the novelist exposes how electricity represents coexisting oppressive and progressive potentialities, but he also illustrates the enduring allure of this energy by composing some of the most electrifying scenes in American literary history.

In my conclusion, I consider what the aforementioned history might mean to readers in the twenty-first century. I discuss the historical and ongoing role that metonymy plays in our representations and conceptualizations of electrical systems. I suggest that the blended study of literature and technology can help scholars explore different facets of the "sociotechnical imaginaries" that shape a society's hopes for its future (Jasanoff and Kim 2009; Jasanoff 2015; Hilgartner 2015). And I reflect upon what happens when a technology becomes implicated with other abstract concepts, such as humanity, embodiment, and modernity.

In each chapter, I select narratives not because they are representative but because they are evocative and formative.[22] To put it another way, the texts I discuss are not time capsules that have preserved the culture in which they emerged; they are thoughtfully composed meditations on human life that will help us ask new questions about our past and about our enduring fascination with electricity and all this energy came to represent. To account for the uniqueness of my cases, I use the familiar methodology of the qualitative case study in a potentially unfamiliar way. Whereas a sociologist or an intellectual historian might utilize a case study as a statistically significant sample of a common trend,[23] I focus on examples that were and are exceptionally thought provoking. Some of my cases are outliers, but each expands upon lively conversations in literary and technological history and each promotes reflection upon our enduring fascination with the electric network and its analogues. Mark Twain, Thomas Edison, Charlotte Perkins Gilman, Charles Steinmetz, Ralph Ellison, and Lewis Mumford all composed narratives that modeled, with rhetorical savvy and sociological perceptiveness, a range of relationships that Americans cultivated with electrical devices and systems. Read as a usable history, their works have the potential to alter how we comprehend technology today even though most of the texts I investigate pre-date the adoption of the word *technology* in the American vernacular.[24]

These texts also complicate how we understand important movements in American literature. By shifting between figurative invocations of electricity and plausible descriptions of actual machinery, they disrupt the boundaries that customarily divide the realistic and the romantic, the historical and the fantastical. As the scholar of science and technology Rosalind Williams argues, "In a world of railroads, telegraphs, electric lights, ether, and x-rays, the marvelous mingled with everyday life. These new experiences recalibrated the literary triad of myth, romance, and realism" (2013, 32–33). The landscape of American literary and cultural history, much like the landscape of American cities, looks different under the glow and the accompanying shadows cast by electric light.

1 MARK TWAIN AND THE TECHNOLOGICAL FALLACY

For many readers, a study of literature and electricity in American history will call Mark Twain to mind. One may have even seen the photograph that was taken of him in Nikola Tesla's laboratory—an image that positions Twain at the nexus of literary and technological history. Twain's interest in invention has become a part of his legend. Guided tours of his Hartford estate now pause in his billiard room to discuss his telephone—the first privately owned telephone in the state of Connecticut. This gadget has come to signify the writer's eccentricity and cleverness. As a great American novelist who also held patents, Twain was distinguished by his passion for technology. He was also bankrupted by it. He lost almost everything after imprudently investing in the Paige Compositor, a typesetting machine with more than 18,000 moveable parts. We know that he hurried to write *A Connecticut Yankee in King Arthur's Court* to pull himself out of his mounting debt; in the novel's preface he apologizes for having rushed the job.

A Connecticut Yankee solidifies the association between Twain and technology in many readers' minds. This novel recounts the experiences of Hank Morgan, an accidental time traveler who re-invents the trappings of nineteenth-century America in sixth-century Camelot.[1] To better understand this character's shortcomings and megalomaniac aspirations, scholars have woven the legend of Twain's deleterious attraction to technology into their readings of the flawed and ambitious text. Introductions to recent editions of *A Connecticut Yankee* inevitably mention the Paige Typesetting Machine and entreat readers to imagine how Twain might have felt about "technological progress" in light of his frustrations with this venture.[2] Generations of literary critics, from James Cox (1960) through Joe B. Fulton (1997), have examined how the author's oscillating

Figure 1.1
Mark Twain holding a wirelessly electrified light bulb in Nikola Tesla's laboratory (originally published in T. C. Martin, "Tesla's Oscillator and Other Inventions," *Century Magazine*, April 1895).

fascination and exasperation with technology plays out in the novel's depictions of advanced weaponry, such as dynamite and electric fences.[3] Even studies that focus on the work's more sustained investment in political economy, humanitarianism, and humor often touch on this theme. It has become nearly impossible to discuss *A Connecticut Yankee* without mentioning technology. Yet the word *technology* doesn't appear in the novel; it was not a word that Twain had at his disposal. As Eric Schatzberg (2006), Leo Marx (1997), and other scholars have demonstrated, the term was not adopted into the American vernacular until years after Twain's death.[4]

In this chapter, I examine how this insight from technology studies might inform a new understanding of this familiar text. Specifically, I argue that literary critics who study *technology* in Twain's novel (and, by extension, in other novels of the same period) have projected an anachronistic coherence onto a variety of images and themes that could not fit so neatly together. The various inventions that the Yankee re-produces—including the bicycle, the telephone, the printing press, the Gatling gun, and the dynamo—could not be described by a single moniker. Whereas late-twentieth-century writers could blame technology for social ills or cite it as evidence for social progress, Twain and his contemporaries would not have formulated such causal constructions. Even if turn-of-the-century writers affiliated machines or inventions with progressive notions of history, they necessarily described these new developments as plural and multifaceted.

This difference in terminology might explain, in part, a discrepancy between historical and present-day readings of *A Connecticut Yankee*. Twain's peers, including the influential editor and writer William Dean Howells, considered the novel to be a "glorious gospel of equality" (Howells 1910). Nineteenth-century readers even regarded the intensely violent "Battle of the Sand-belt" episode—in which Hank repurposes his "civilizing" inventions to electrocute, explode, and drown 25,000 knights—as a humane and realistic depiction of modern war. Eight years after the novel's publication, a writer for the trade journal *The Electrical Age* praised *A Connecticut Yankee*'s depiction of advanced weaponry, projecting "that future warfare ... will be built on some such lines" ("Electricity in War," April 24, 1897, 263).[5] As late as 1941, *A Connecticut Yankee* was included on the *Journal of Educational Sociology*'s "Reading List for Democracy" with an unambiguously positive description: "Story of a mechanically minded Yankee transported backwards in time to the England of King Arthur's time, expressing an idea of freedom."

Darker interpretations of the novel gained prominence in the 1960s, as Henry Nash Smith (1964), Sherwood Cummings (1960), and other notable scholars introduced the word *technology* into their criticism, along with its postwar connotations.[6] During those years, *technology* began to elicit fear—not only because of the Cold War threat of nuclear disaster, but also because the word itself had become a scapegoat for diffuse cultural anxieties. As a difficult-to-define singular noun, *technology* could sneak insidiously into the subject of sentences. *It* could be dangerous or beneficial to humanity.

This relatively recent tendency to use *technology* as a singular term seems unique to the English language. As Schatzberg argues (2006, 489), "Whereas a single term seems adequate in English, Continental languages use two, the cognates of *technique* and *technology*." Schatzberg adds that, after 1940, in English "the term no longer referred to the study of material culture, but rather to material culture itself" (493). According to Thomas Misa (2003, 9), reification of this concept has "overaggregated" *technology* into an abstract and deceptively coherent term. In the process, it has also promoted despair about the modern condition by allowing material artifacts and systems to seem like an unified, extrinsic force that affects users against their wills.

As *technology* evolved into the singular concept we now know, it also came to appear timeless. Perceiving *technology* as pervasive and ahistorical, even the most careful scholars have projected this term onto the past as if it had always existed everywhere.[7] Hence, Twain's novel can appear to be about *technology* although it doesn't contain any occurrences of that word. Leo Marx rationalized this tendency to project *technology* onto the past by suggesting that the idea existed before there was an "adequate concept" to denote it: in an elegant turn of phrase, he argues that the word emerged to fill a "semantic void" (Marx 1997, 967). Although artifacts that we would now identify as technological existed before that term came into popular use, I disagree with Marx on this point. I call these acts of simplification and projection *technological fallacies* because they function much like William K. Wimsatt and Monroe C. Beardsley's (1946) *intentional fallacy*, shaping the way we approach literary and cultural history.[8]

The concept *technology* didn't emerge in the American vernacular with a precise, useful definition; instead it accumulated connotations in feedback with practices, institutions, and social mores.[9] As Schatzberg argues (2006, 487), "When the term [technology] did become widespread in elite discourse in the United States, it bore the stamp of a long struggle over the meanings of industrialization." Furthermore, what existed before the word *technology* garnered its present meanings was not a void. To describe technical developments, Twain and his peers employed diverse narrative strategies, such as drawing analogies to electric power. When allusions to electricity weren't precise enough, they crafted hybrid images that complicated their analogies to this energy. Just as electrical systems linked various components into a circuit to emit light or heat, Twain and other *fin de siècle* writers linked disparate images to create varied rhetorical effects. These narrative strategies permeate *A Connecticut Yankee* (1889), and they are the focus of this chapter.

For example, when the Yankee recognizes his opportunity to re-create (and thereby control) "civilization," he describes his endeavor by concocting a hybrid electrical metaphor. He begins by comparing his re-inventions to a volcano, an emblem of the natural sublime[10]: "Unsuspected by this dark land, I had the civilization of the nineteenth century booming under its very nose! … There it was, as sure a fact, and as substantial a fact as any serene volcano, standing innocent with its smokeless summit in the blue sky and giving no sign of the rising hell in its bowels" (120).[11] The volcano calls forth the raw, devastating power that Hank sees in the systems he re-creates, but this image cannot capture his sense of ownership. He layers in a new metaphor: "I stood with my hand on the cock, so to speak, ready to turn it on and flood the midnight world with light at any moment. But I was not going to do the thing in that sudden way. It was not my policy" (120). Hank describes civilization (or the systems that constitute it) as part volcano, part blinding electric light. Twain's vision is too scientifically advanced and carefully planned to be represented by an unpredictable natural disaster; its potential for devastation is too dramatic to be signified by a light switch. He must knit together analogies to electrical power and a more ancient symbol of sublimity to evoke the complexity of the modern systems he rebuilds.

Twain also employs electricity as a metaphor in its own right. In fact, he dithers between treating it as a symbol and as a material power source. In the tenth chapter of *A Connecticut Yankee*, "Beginnings of Civilization," he omits electricity from the list of "conveniences" that his protagonist longs for and eventually re-produces. This absence is conspicuous in view of the fact that the chapter opens with an illustration of an electrician stringing power cables—and considering that Twain added this chapter near the end of his writing process, after he had already established the prominent role that electric power would play in the text.[12] The Yankee plainly asserts that a country needs a patent office, a newspaper, glass, schools, churches, type-writers, telephones, telegraphs, soap. He even jokingly introduces stove polish before inventing a stove. He never mentions the need for, or construction of, a light and power plant in this chapter or any other. Instead, he subsumes all of his other developments under metaphors that are drawn, at least in part, from this invention. In the paragraph following his hybrid image of the volcano and the light switch, Hank extends the latter metaphor: "I was turning on my light one-candle-power at a time, and meant to continue to do so" (120). In this episode, electrical light is not an invention Hank explicitly builds in the service of re-creating American modernity. It is the essence of his whole industrial enterprise.

Whether or not Twain intended to position electricity as a metonym for modernity, these moments reveal how quickly Americans came to understand electricity as a natural part of their lives. It literally goes without saying that the Yankee's modern version of Camelot will be electrified, although electrical light and power would not be available to most Americans for another forty years. Hank only indicates that he turned actual lights on when he notices that the Church turned them off: "From being the best electric-lighted town in the kingdom and the most like a recumbent sun of anything you ever saw, it was become simply a blot—a blot upon darkness—that is to say, it was darker and solider than the rest of the darkness, and so you could see it a little better; it made me feel as if maybe it was symbolical" (326). Like twenty-first-century Americans who confront their reliance on electricity only during a power failure,[13] Hank was so thoroughly accustomed to electrical light that its absence seems more remarkable than its presence.

The Yankee's reaction to the blackout reveals that he confuses electrical apparatuses with the "symbolical" meanings he projects onto them. Earlier in the novel, he suggests that electrical inventions correspond to his notion of progress. This episode implies that they *constitute* it. Hank reads the absence of electric light as proof that the people have rejected enlightenment, even though he figuratively kept Camelot "in the dark" by single-handedly controlling the means of producing his social and technical systems. In this sense, the theme of electricity functions in the novel much like the term *technology* does in twentieth-century and twenty-first-century discourse: it seems precise, but it has multiple and contradictory meanings. Its illusory ambiguity allows Americans, like those Hank represents, to confuse whether it represents "progress" or inspires, embodies, or inhibits it.

This ambiguousness made electricity an appealing subject for Twain and other contemporary writers. Inherently plural, electricity could

CHAPTER X.

BEGINNINGS OF CIVILIZATION.

THE Round Table soon heard of the challenge, and of course it was a good deal discussed, for such things interested the boys. The king thought I ought now to set forth in quest of adventures, so that I might gain renown and be the more worthy to meet Sir Sagramor when the several years should have rolled away. I excused myself for the present; I said it would take me three or four years yet to get things well fixed up and going smoothly; then I should be ready; all the chances were that at the end of that time Sir Sagramor would still be out grailing, so no valuable time would be lost by the postponement; I should then have been in office six or seven years, and I believed my system and machinery would be so well developed that I could take a holiday without its working any harm.

I was pretty well satisfied with what I had already accomplished. In various quiet nooks and corners I had the beginnings of all sorts of industries under way—nuclei of future vast factories, the iron and steel missionaries of my future civilization. In these

Figure 1.2
"Beginnings of Civilization" (illustration by Daniel C. Beard in first edition of *A Connecticut Yankee in King Arthur's Court*).

connote the life force of "animal magnetism," the death spark of the electric chair, the recent discoveries of modern physics, the corporations that made it commercially available, and much more. In *A Connecticut Yankee*, it appears as a lightning bolt with mythic connotations, as actual and metaphorical electrical buttons, and as dynamo-generated power. Indeed, some of the crucial tensions in the novel—between magic and science, romance and realism, hierarchy and democracy—emerge when Hank muddles these multiple meanings of electricity.[14] His uses of electricity dramatize the complications that arise when one broad concept operates on various symbolic and realistic levels. Much of this subtlety has been overlooked by studies that fallaciously conflate Twain's depictions of electricity with other *technologies*, especially including the telephone.

If the novel's various representations of electricity appear divergent, note that *A Connecticut Yankee* describes its own contents as a "palimpsest"—a parchment that has been repeatedly overwritten, leaving a perceptible impression of each of the lines that came before it. By using this organizing image in his introduction and conclusion, Twain invites readers to track how meanings accumulate across episodes and throughout history. I accept his invitation by reading the novel's treatment of electricity as one dimension of its palimpsest-style storytelling. Beginning with Twain's most surrealistic depictions, I chronicle the "effects" of electricity as it appears in various forms throughout the novel. Each depiction of electricity, as an analogy, a tool, and an aspect of American history, contributes a layer to the story. Twain's playful imagery remains imprinted upon the novel's electrocution scenes; the trace of the macabre lends depth to Twain's surrealistic, "shocking" jokes. By examining these layered meanings, I illustrate how electricity became a complex, allusive symbol of modernity for Twain's generation. I also demonstrate how much we might learn about this novel—and perhaps about our own cultural moment—by disaggregating the artifacts, systems, and symbols that we collapse into the term *technology*.

LIGHTNING AND OTHER "SYMBOLICAL" SHOCKS

Electricity played a unique role in Twain's thinking about Camelot from the start. He dreamt up the basic literary burlesque of *A Connecticut Yankee* in December of 1884, shortly after reading *Morte D'Arthur* at George

Washington Cable's recommendation. An 1884 entry from Mark Twain's personal journal—written shortly before his entry about Detroit's electric light—captures his first idea for the novel. Picturing his own body underneath a suit of armor, he pokes fun at the simple, human urges that romance writers omit from their knights' tales: the need to urinate, to itch, to sniffle. Only his mention of lightning deviates from this comic premise of describing a knight realistically.[15] Where his other jokes draw attention to the fleshy reality of the human body, "Always getting struck by lightning" suppresses the affect an electric shock would have on an actual person:

> **X** Dream of being a knight errant in armor in the middle ages.
>
> **X** Have the notions & habits of thought of the present day mixed with the necessities of that. No pockets in the armor. No way to manage certain requirements of nature. Can't scratch. Cold in the head—can't blow—can't get at handkerchief, can't use iron sleeve. Iron gets red hot in the sun—leaks in the rain, gets white with frost & freezes me solid in winter. Suffer from lice & fleas. Make disagreeable clatter when I enter church. Can't dress or undress myself. Always getting struck by lightning. Fall down, can't get up. See Morte Darthur. (Twain 1979, 78)

The comedic value of the electric shock amused Twain, and no wonder. Like the form of burlesque, this image transforms serious content into silliness.[16] It generates humor from the uncomfortable truth that electricity transforms the human body into a conductor. Twain returns to this fact both jokingly and seriously in *A Connecticut Yankee*.

The "Always getting struck by lightning" gambit doesn't appear in the published version of the novel. Twain opted to have his protagonist control lightning rather than comically succumb to it. Yet he doesn't abandon this imagery entirely. In the episode based upon this journal entry, "The Yankee in Search of Adventures" (chapter 11), Twain repurposes this joke into two analogies. The first describes the protagonist's experience of donning armor for the first time. Echoing Twain's journal entry, the passage parodies the romantic notion of the chivalric knight by ridiculing the physical discomfort of his armor: "all the while you do feel so strange and stuffy and like somebody else—like somebody that has been married on a sudden, or struck by lightning, or something like that, and hasn't quite fetched around, yet, and is sort of numb, and can't just get his bearings" (136). Much like the image of the volcano and the lightning switch one chapter earlier, this incongruous pairing emphasizes Hank's struggle to articulate thoughts that pull him in multiple directions. In this passage, he elaborates upon the dead metaphor "thunderstruck" with unprecedented detail, correlating a lightning strike with intense disorientation akin to the figurative shock that Twain refers to as "electric surprise" (18) in the novel's frame narrative. By weaving in a bizarre analogy to an unanticipated marriage, Hank underscores his sense of disassociation from himself. At his best, this character uses his modern vocabulary just as he uses other recent inventions: to demonstrate his superiority over his sixth-century companions. But this compound simile marks the limitations of that lexicon. When Hank lacks the words to describe the modern or the ancient world, he cobbles together hybrids that evoke complexity through incongruous juxtaposition.

Hank's next analogy amplifies the morbid undertones of the electric-shock trope. Trying to nap in his suit of armor, he feels cold, first, and insects crawling across his body, second. "Even after I was frozen solid," he complains, "I could still distinguish that tickling, just as a corpse does when he is taking electric treatment" (153). This passage extends the burlesque Twain originally imagined in his journal entry, mocking the impracticality of armor by drawing a comparison that seems ridiculous to nineteenth-century sensibilities. The "corpse … taking electric treatment" playfully alludes to Luigi Galvani's experiments, which determined how muscle fibers react to electrical currents. This reference might also call to mind gothic and humorous representations of these experiments.[17] Once again, Twain's treatment of electrical power draws focus from the uncomfortable human body under the armor to a more surrealistic image with concurrent morbid and humorous implications.

The latent dangers of electricity give these jokes their edge. They also give Hank an edge over Merlin when they compete as magicians. After being arrested and sentenced to death for trespassing in the opening sequence of the novel, Hank pretends to control a fortuitous solar eclipse to intimidate his captors into freeing him. With his audience already cowering, he appeals to ancient superstitions about lightning to accentuate the effect: "If any man moves—even the king—before I give him leave, I will blast him with thunder, I will consume him with lightnings!" (75). His threat capitalizes on the fear of lightning as a devastating natural force and as a mythic symbol of punishment from the gods. He inflates and lampoons both fears in the following episode (chapter 7), when he harnesses lightning to explode Merlin's Tower and secure his position as the best magician in Camelot.

Throughout the Tower episode, Hank manipulates multiple meanings of lightning as fearsome and funny, mythical and understandable through science, in order to entertain on multiple levels simultaneously. For his sixth-century spectators, he exploits the supernatural connotations of lightning, heightening the awesome effect of the bolt by drawing (and apparently controlling) its charge and by exaggerating its effects with a hidden cache of gunpowder. For his nineteenth-century readers, he plays up the scientific applications of electricity, emphasizing his senses of control and of humor. Hank's layered narration extends the novel's core burlesque by poking fun at his spectators' naiveté—especially Merlin's. It also garners credibility for the Yankee, aligning him with nineteenth-century electrical experts who used humor to distinguish themselves from laypeople.[18]

Hank sets the stage for his performance by letting his nineteenth-century reader peek behind the proverbial curtain: "When the thirteenth night was come we put up our lightning-rod, bedded it in one of the batches of powder, and ran wires from it to the other batches. Everybody had shunned that locality from the day of my proclamation, but on the morning of the fourteenth I thought best to warn the people, through the heralds, to keep clear away—a quarter of a mile away" (88). This passage constructs the novel's readers as insiders who grasp how the electrical "miracle" works. It establishes Hank's careful forethought in staging the spectacle and in guarding the public from potential harm. It also characterizes electricity as an easily comprehensible and safely useable energy, reinforcing the conceit that "modern science" has "tamed" this unwieldy natural power.[19]

After Hank sets up his simple circuit, he waits. Days later, at the first sign of a storm, he begins his performance. The show begins with an empty gesture of showmanship. He calls out to Merlin: "You wanted to burn me alive when I had not done you any harm, and latterly you have been trying to injure my professional reputation. Therefore I am going to call down fire and blow up your tower, but it is only fair to give you a chance" (49). Using "magic" instead of a lightning rod, Merlin fails. Hank takes the center stage, waiving his arms and proclaiming: "You have had time enough. I have given you every advantage, and not interfered. It is plain your magic is weak. It is only fair that I begin now" (49–50). Shortly thereafter, lightning strikes.

Hank's lightning rod circuit harnesses atmospheric electricity to create a "volcanic fountain of fire," foreshadowing the hybrid metaphor of the volcano and the light switch he will introduce three chapters later in "The Beginnings of Civilization." The chapter "Merlin's Tower" generates tension by invoking deep fears about the destructive capacity of electricity, and—in the manner of Twain's jokes—it generates humor by discharging this tension with comedic misdirection. As the culture-industry theorists Max Horkheimer and Theodor Adorno explain (2002, 112), "laughter, whether reconciled or terrible, always accompanies the moment when fear is ended." While this passage ostensibly focuses on a lightning strike, it draws attention away from the lightning itself by emphasizing the amusing discrepancy between Merlin's ignorance and Hank's mastery. Hank interjects an economic metaphor, "Merlin's stock was flat," to mark this transition from sublimity into humor—and perhaps to align his performance with the economy of sensational entertainment that was emerging at the end of the nineteenth century.[20]

Twain underscores the Yankee's dominance over Merlin by focusing on his narrator's body. Privy to his behind-the-scene circuits, the reader recognizes that Hank's arm waving doesn't affect the explosion. Hank even confides that he takes his cue from the lightning itself, since he only begins his act after his lightning rod "would be loading itself." Still, by drawing out the description of his circuitry and his showmanship, the narrator insists that his readers and spectators recognize his power as an electrical expert and as a magician, respectively. By detailing how Hank manipulates the lightning's charge (by channeling it into gunpowder) and its mythology (by staging this performance), this episode begs the question: is the destructive potential in the lightning itself, or in the Yankee's exultation in controlling it?[21]

Thinly veiled allusions to Benjamin Franklin raise similar concerns about Hank's character. The narrator never mentions him by name, but Hank conjures Franklin's image by building a lightning rod while aspiring to build a nation.[22] As James Delbourgo argues, widely circulated and reproduced works such as Marguerite Gérard's *Au Génie de Franklin* linked Franklin's electrical and political exploits, portraying Franklin as "a Prometheus/Jupiter figure whose godlike mastery of nature allow[ed] him to overcome British enemies with electrical lightning bolts" (Delbourgo 2006, 4). In view of the popularity of this legend, contemporary readers would be likely to recognize a parallel between Franklin and the Yankee in this scene. This resemblance lends to Hank's ethical appeal. But, as with any analogy, it also invites readers to consider the insufficiencies of the comparison.

Considering the prestige that Franklin earned with his lightning rod in the eighteenth century, readers might surmise that the Yankee could have secured sixth-century Camelot's esteem with this re-invention alone. Had he done so, Hank would have left Merlin's tower intact. Yet the Yankee improvises as he plays Franklin, deviating appreciably from his predecessor's example. Apparently enchanted by devices that summon power with the touch of a button, Hank finds the lightning rod passive. Even when it can draw a charge, he perceives the act of attracting lightning as mundane in itself. He opts to discharge the lightning into gunpowder to make the spectacle suitably impressive. In so doing, Hank insinuates that electrical phenomena are most interesting when he can control or enhance them. Unlike Franklin, who was fascinated with the nature of electricity, Hank cares little for the mystery of this physical phenomenon. He prefers to use it. This subtle change in motivation transforms lightning into volcanic fire—a dangerous bit of alchemy.

Whether readers find Hank's spectacles scintillating or frightening, his deviation from Franklin's legend registers a change in the way Americans understood electricity. Electricity represented a natural wonder to the Enlightenment-era scientist and poet; to the *fin de siècle* American, it represented many things at once. In Twain's chapter titled "Merlin's Tower," electricity appears practically useful and entertaining. It also appears destructive. Hank recounts how the bolt—with a bit of forethought and planning—obliterates Merlin's tower: "there was an awful crash and that old tower leaped into the sky in chunks, along with a vast volcanic fountain of fire that turned night to noonday, and showed a thousand acres of human beings groveling on the ground in a general collapse of consternation" (90–91). Breaking from the rational tone with which Twain described Hank's behind-the-scenes wiring, this rapturous run-on sentence likens the effect of the lightning circuit to a volcanic eruption.

Lightning might not mystify a nineteenth-century American as it did an eighteenth-century one, but it still evokes awe. In this passage, electricity appears to be more tangible and potentially destructive than it seems in Twain's jokes or in his metaphorical language. Yet if this explosive scene arouses any anxieties in the reader, Hank quickly diffuses the tension. He changes his tone a third time, reframing the episode as a joke: "Well, it rained mortar and masonry the rest of the week. This was the report; but probably the facts would have modified it" (91). This observation reminds readers that Hank has designed this performance to ridicule the outsiders who irrationally trust gossip instead of "facts."

Hank ultimately uses three narrative modes in the lightning rod episode. He describes his electric circuit first as a scientific demonstration, then as a sublime display of force, and finally as a harmless joke played on rubes who don't understand the useful physical properties of lightning. By weaving together the scientific, the sublime, and the humorous, Hank produces a composite image that exploits the layered significations of lightning. Although this heterogeneous depiction is more elaborate than his analogy to the volcano and the light switch, the narrative strategies Hank uses in these examples resonate with one another. He cannot describe electrical phenomena or "civilization" plainly; he creates hybrid images every time that he attempts to describe either.

The note on which Hank closes the "Merlin's Tower" episode sets the tone for the majority of the novel. By rebuilding the tower at the end of the chapter, he emphasizes his own good-natured spirit and erases the evidence of the lightning rod's destructive potential. By ensuring that his electrical display will leave no lasting mark on Camelot, Hank departs further from Franklin's example. Whereas Franklin's lightning rod became permanently affixed on the rooftops of buildings and in narratives about American history,[23] Hank's lightning rod disappears entirely. After this scene, he never mentions it again. The contrast between Franklin's legacy and the Yankee's indicates the endlessly replaceable nature of electrical invention from the latter's perspective. The same lightning rod that made Franklin seem godlike in the early days of the American republic cannot hold the attention of the nineteenth-century American. Since Hank understands his own inventions to be as fleeting as the lightning strike itself, he believes that he will have to prove himself repeatedly. To do so, he will turn to electricity almost every time.

THE BATTERY-POWERED BUTTON

Hank's fascination with electrical power is unique in the novel. While other artifacts we now would classify as technologies—ranging from bicycles to dynamite—also attract Hank's inventive attention, electricity is the only invention that this character describes as a hybrid. He figuratively associates it with volcanoes, literally links it with gunpowder, and he uses a blend of realistic and romantic conventions to describe its stunning "effects." Provocatively, Hank doesn't describe telephones and telegraphs with similar strategies, although these systems use electricity. His communication systems are not sublime. They are ironic; Hank's systems disconnect people while purporting to connect them.

Although Hank claims that he intends his telephone and telegraph systems to interconnect (and thereby "resurrect") the nation, he forces his workers to detach themselves from society entirely to accomplish this goal. Hank explains: "My men had orders to strike across country, avoiding roads, and establishing connection with any considerable towns whose lights betrayed their presence" (121). This passage suggests that Hank's networks connect a series of lights rather than people. Twain later underscores this theme by revealing that the telephone and telegraph operators are removed from society entirely. When Hank happens upon a newly erected telephone station, he is disappointed to learn that the operator has not heard of the "miracle" he performed only feet away. The operator explains: "Ah, ye will remember we move by night, and avoid speech with all. We learn naught but that we get by the telephone from Camelot" (181).

By forcing this schism between his telephone workers and the greater public, Hank creates a new social fault line that hinders his attempt to close the gap between the aristocracy and lower classes. The Yankee had hoped to democratize knowledge with this invention, but his private telephone system redirects and skews it. This depiction anticipates social formations that network theorists have discussed in recent usage pertaining to the Internet and to smartphones. In Barry Wellman's terminology (2002, 10–25), Hank uses the telephone to replace "place-to-place" communications with "person-to-person" communications. In so doing, he disperses and fragments Camelot's communities by removing his operatives from their homes and forcing them to deal with other people exclusively through his developing systems.[24]

Consider, in contrast, Hank's second electrical spectacle—an episode that coincides with his second encounter with Merlin. In the fifteen intervening chapters between showdowns, the protagonist initiated what he called the "beginnings of civilization," constructing his figurative volcano and light switch (chapter 10). He earned the moniker "The Boss."[25] He also played the knight in shining armor, wandering the countryside to aid the damsel in distress, Alisande ("Sandy"). Hank became a sympathetic character in these interludes. He freed enslaved people and lamented the cruelty of capital punishment and inequitable economic systems. Despite his ostensible growth, he reverts to his power-hungry antics when he discovers Merlin trying to fix a broken well in the Valley of Holiness. Immediately, Hank decides to use his practical knowledge about electricity and gunpowder to assert his superiority once again.

As in the "Merlin's Tower" chapter, this two-chapter episode (chapters 22 and 23) combines a backstage description of circuitry with a staged show. The "Restoration of the Fountain" begins after Hank learns that a holy well has gone dry and that the resident monks interpret the problem as a punishment from God. The Boss wanders onto the scene after the monks have asked Merlin to remedy the well with magic. Hank scoffs at this superstitious solution. Believing the well to have sprung a simple leak, he secretly investigates the problem with several students he recruited in intervening episodes. The group easily mends the well, but Hank wants more than appreciation for a repair job well done. To impress the crowd and disgrace Merlin, he builds another circuit that conducts an electric charge into gunpowder. Again, he describes his backstage setup with technical precision: "We grounded the wire of a pocket electrical battery in that powder, we placed a whole magazine of Greek fire on each corner of the roof—blue on one corner, green on another, red on another, and purple on the last—and grounded a wire in each" (289). This time, instead of waiting for lightning, Hank uses an electrical battery and a button to control the effect more completely. This choice to use a button warrants particular consideration because electrical buttons were relatively new and were still feared by some of Twain's contemporaries at the time of the novel's publication.[26] This episode thus engineers the over-reaction of the sixth-century audience in order to poke fun at nineteenth-century Americans who continue to fear electricity.

The button also demands attention because Hank didn't have to use it. In his previous contest with Merlin, our hero put atmospheric electricity on display. In this episode, electricity is entirely ancillary. The Boss admits that he could have impressed his spectators by merely fixing the well. He could have "enlightened" them by showing them how he had done it. He even admits that he could have astounded them with the relatively unsophisticated water pump he devised, boasting that this mechanism "was a good deal of a miracle itself" and that the monks "were full of wonder over it" (295). Hank doesn't need electricity to fix the fountain. He explains that it simply adds "style":

> When you are going to do a miracle for an ignorant race, you want to get in every detail that will count; you want to make all the properties impressive to the public eye; you want to make matters comfortable for your head guest; then you can turn yourself loose and play your effects for all they are worth. I know the value of these things, for I know human nature. You can't throw too much style into a miracle. It costs trouble, and work, and sometimes money; but it pays in the end. Well, we brought the wires to the ground at the chapel, and then brought them under the ground to the platform, and hid the batteries there. (288–289)

The philosophy of showmanship Hank espouses in this episode emphasizes spectacle over substance. In the Franklinian tradition, experts staged electrical displays in order to educate and to awe, but Hank aspires to accomplish only the latter.[27] Shifting his attention from the public good to "the public eye," he re-codes electricity as a convenient medium for entertainment. In so doing, he reveals his allegiance to a "new public culture" that the scholar of American studies Lauren Rabinovitz (2012, 2) describes as "dedicated to the consumption of enjoyment"—a culture that apprehended "disaster shows" as entertaining rather than horrifying.[28]

With the circuitry built, Hank begins his sixth-century spectacle with speeches and incantations. Then he presses his electric buttons. As in the "Merlin's Tower" episode, Hank shifts from recounting his circuit matter-of-factly to describing its effect exuberantly: "I touched off one of my electric connections, and all that murky world of people stood revealed in a hideous blue glare! It was immense—that effect! Lots of people shrieked, women curled up and quit in every direction, foundlings collapsed by platoons. The abbot and the monks crossed themselves nimbly and their lips fluttered with agitated prayers" (291).[29] Earlier, Hank described his civilization as volcano-like. Now, he emulates the explosive force of a real volcano, using his hidden circuits to ignite fire bursts. His audience reacts as if they are observing a sublime natural phenomenon.

Hank's marriage of the electrical and sublime imagery situates this scene in the tradition that scholars today might call the "technological sublime."[30] Nye (1994, xiii) describes this emotional response as "one of the most powerful human emotions," adding that "when [it is] experienced by large groups the sublime can weld society together." Read in this context, Hank's divergent ambitions to champion democracy and "to impress an ignorant race" seem to converge: by staging the sublime, Hank might elicit a feeling of awe and unity akin to that evoked by fireworks during present-day celebrations.

At the same time, Hank's delight in his viewer's supplication reveals his capacity for destruction. Indeed, this episode exaggerates the protagonist's perversity. Unlike the "Merlin's Tower" episode, which ends with jokes that diffuse the tension, this episode ends with self-congratulation: "It was a great night, an immense night. There was reputation in it. I could hardly get to sleep for glorying over it" (295). Though this "miracle" was constructive rather than destructive, Hank's boastfulness displaces his warm sense of humor, and it subtly alters the novel's depiction of electricity. Earlier, this energy represented magic, science, and a somewhat patronizing sense of humor. In this chapter, electricity signifies magic, science, and Hank's unquenchable thirst for attention.

Although "The Restoration of the Fountain" lacks the lighthearted dimension of "Merlin's Tower," both episodes blend realism and sensationalism in their depiction of electrical phenomena. The consistent use of this hybrid rhetorical technique hints at the complex characterization of electricity at the time of the novel's publication. To Hank and to his readers, electricity acts both as a usable energy that behaves according to predictable physical laws and as a reminder of the invisible and surprising aspects of the universe that humans were just beginning to discover. This heterogeneous characterization of electricity might explain why Hank opts to use this energy to ignite his fireworks, rather than a simple match and fuse: electricity appears both numinous and practical in the American cultural imagination.[31] It enables the Yankee to act as an expert and a magician simultaneously. And, considering Twain's familiarity with contemporary constructions of the nervous system,[32] the authorial decision to use electricity creates an evocative parallel. As Hank presses buttons to pulse electrical power through a wire, Twain manipulates the electrical impulses of his reader's imagination. Both the author and his protagonist use electricity to create a shock.

THE ELECTRICAL SYSTEM

In the "Merlin's Tower" and "Restoration of the Fountain" episodes, Hank connects electricity to gunpowder, indicating a correspondence between these inventions in his imagination. Both can produce beautiful effects, such as fireworks and light. Both also can destroy just about anything. Before Twain explores the fatal capacities of this combination in the climactic "Battle of the Sand-belt" (chapter 43), he briefly severs this link between gunpowder and electricity in "The Yankee's Fight with the Knights" (chapter 39) and "Three Years Later" (chapter 40). The "Fight" episode features the Yankee's third contest with Merlin, in this case by proxy. In this chapter, the Boss faces off against Sir Sagramor, a knight whose armor has been protected by Merlin's spells. For a third time, Hank's "magic" wins. He triumphs by forgoing a knights' armor and arms in favor of his own inventions: comfortable, form-fitting clothing and a gun.

Whereas Hank's earlier showdowns channeled electricity into gunpowder to perform harmless spectacles, this episode employs gunpowder alone—this time to kill. As the literary scholar Jane Gardiner notes (1987, 450), the central confrontation of this chapter is unique for its lack of electrical imagery: "Only the defeat of Sir Sagramor, with a bullet from a revolver, is secured by some means unconnected with electrical power." In many respects, this contest parallels the Yankee's electrical spectacles. Once again, Hank uses his practical knowledge to embarrass Merlin and to astound his audience. Once again, his spectators cannot perceive how Hank managed his miracle—both the bullet and the electric current travel quickly enough to elude the human eye. Despite these similarities, Hank's narrative style changes with his choice of invention. He recounts this victory dispassionately, without the relish of his earlier performances. He remarks, nonchalantly, "I snatched a dragoon revolver out of my holster, there was a flash and a roar, and the revolver was back in the holster before anybody could tell what had happened" (506).

Why does Hank depart from his pattern of showmanship in this chapter? I can think of two likely answers to this question. Since Hank has emphasized the ephemerality of electrical spectacles throughout the novel, he might use a gun here to make a more lasting impression. Indeed, in the "Word of Explanation" that precedes the novel the frame's narrator, Mark Twain, mentions the mystery of the "round hole through the chainmail" on display in Warwick Castle. Thus, the use of gunpowder without electricity enables Hank to leave a mark that Twain can use to corroborate Hank's story. Another more nuanced answer to this question might involve the phenomena he attempts to describe. Hank uses layered narrative styles and hybrid analogies to describe electricity, suggesting that this energy may confound linear narration in a way that gunpowder does not. Indeed, as I will discuss in the next chapter, the rhetorical appeal of electricity lies in the perceived discrepancy between cause and effect: the straightforward arrangement of wires and slight touch of Hank's finger can create figurative volcanoes and it can bring a whole crowd to its knees. To describe only the cause *or* the effect would be to diminish this astounding contrast. Guns are similar in that they, too, require only the pressure of a finger to operate. But electricity is imperceptibly silent whereas guns are loud. And *fin de siècle* American literature had established conventions for describing gun violence, while the conventions for describing electricity still were evolving. Thus, Hank doesn't have to weave together different narrative conventions to describe his gunshot; a quick mention of its "roar" suffices.

This change in Hank's narrative style might also allow him to explore the distinct connotations of gunpowder and electricity. When he isolates electricity from gunpowder, he depicts the latter as dangerous and the former as helpful. By temporarily disentangling the association between gunpowder and electricity, Twain sets the stage for "Three Years Later," the false denouement to the novel. That chapter begins with Hank consolidating his victorious "fight with the knights" by posting a standing threat for all to see: "[N]ame the day, and I would take fifty assistants and stand up *against the massed chivalry of the whole earth and destroy it*" (512).

Here the ambiguous pronoun *it* refers either to "the whole earth" or to the "massed chivalry," underscoring the severity of Hank's warning and foreshadowing the mass destruction he will unleash in the Battle of the Sand-belt. The Boss justifies this tactic paternalistically, insinuating that he threatens Camelot for its own good. For some reason, he believes that this threat enables him to complete the plan he started in the chapter titled "Beginnings of Civilization." After posting his warning, he boasts: "I exposed my hidden schools, my mines, and my vast system of clandestine factories and work-shops to an astonished world. That is to say, I exposed the nineteenth century to the inspection of the sixth" (511).

As other scholars have argued, the Yankee's hankering for violence inheres in this "inspection" process, despite his assurances to the contrary.[33] Noting that Hank begins preparing "to send out an expedition to discover America," critics have interpreted the "Three Years Later" episode as a satire of American imperialism—an issue that seems to have preoccupied Twain in the final decades of his career. For example, Robert Shulman has argued (1987, 151) that Hank's cognitive dissonance maps onto longstanding political and economic tensions: "The concealed tensions that have plagued American republicanism from the days of Franklin and [James] Madison reappear in the conflict between Hank's commitment to liberty and freedom and his own impulse to dominate others and become Boss, a conflict rooted in the market society and in Hank's character and related to his seriously divided feelings about the common man he is ostensibly trying to save." I agree, but I add that these tensions become affiliated with Hank's inventions in significant ways. Throughout this episode, Hank separates his penchant for destruction from his inventions. With this subtle modification, he diminishes the novel's earlier association between electricity and violence.

THE ELECTRICAL SLAVE

In the chapter titled "Three Years Later," Hank characterizes his inventions as singularly humanitarian. He boasts: "Slavery was dead and gone; all men were equal before the law ... and all the thousand willing and handy servants of steam and electricity were working their way into favor" (513). Here Hank characterizes electricity as useful and ostensibly harmless, but the language he uses warrants some consideration. In the terms of nineteenth-century advertisements, Hank describes electricity and steam as "servants." His syntax renders the slave-master relationship ambiguous. It isn't clear whether he describes servants that are *constituted of* steam and electricity or machines that are handy servants *to* steam and electricity. In either case, by also announcing that "Slavery was dead," the narrator implies that electricity and steam facilitated social equality by reducing economic dependency on human slavery. Twain thus plays on the notion of the "electrical slave," a construct that appeared in novels, advertisements, advice columns, and technical journals throughout the nineteenth and twentieth centuries.[34]

The idea of the electrical slave epitomizes the politically charged language that I address in this book. Depending on how it was deployed, this idea could displace attention from the struggles surrounding labor reform by implying that electrical inventions could unproblematically operate as slaves. *A Connecticut Yankee* doesn't go that far. It simply suggests that social progress coincides with the adoption of modern inventions. Industry narratives were less subtle. They re-cast human problems as technical, and they displaced the history of (and need for) social action by implying that new inventions would resolve social issues. The scholar of literature and technology Leo Marx (1997) dubbed "technology" a "hazardous concept" for promoting this type of slippage between the agency of artifacts and the responsibility of individuals or social groups.[35]

Literature could raise questions about this determinism. For example, Hank Morgan's faith in electrical servants may be called into question when Camelot rejects his inventions at the end of the novel. Still, satire is a subtle art, and Twain doesn't explode the logic of the electrical slave or servant by putting this language into the mouth of an unreliable narrator. As I will discuss in chapter 5, the notion of the electrical slave persisted for decades in the advertising character Reddy Kilowatt and in the thousands of industry-generated promotions of a commodity-intensive lifestyle that promised to save labor. In this massive nationwide advertising campaign, "slave" functions mostly as a verb.[36] It represents an action that middle-class Americans can avoid by purchasing the right gadgets, rather than a part of American history with enduring social and economic effects.

Not only does Twain appeal to the controversial notion of the "electrical slave" in the aforementioned episode; he also forces his "helpers" into

Arthurian homes with the threat of violence—arguably compounding the violent connotations of the "servant" metaphor. After he unveils his inventions, the Boss hires the very knights he recently defeated to become his "effective spreaders of civilization." He explains: "They went clothed in steel and equipped with sword and lance and battle axe, and if they couldn't persuade a person to try a sewing machine on the instalment plan ... they removed him and passed on" (513). Insofar as the Yankee self-identifies as an inventor whose dominance follows from his mechanical and scientific savvy, his decision to win his empire by the force of old weapons rather than new ones appears outstandingly uncharacteristic. It seems that Hank wants his readers to set aside for the moment the pernicious capacities of modern inventions, so that they might better appreciate the utopian potential that lies dormant in each of the artifacts and systems (or servants) that he has re-invented.

Since Hank is willing to impose his inventions on an unwilling populace, he ostensibly believes that the very presence of new gadgets will incite the change he hopes to see. He doesn't realize that his co-optation of other forms of violence undermines his stated goal.[37] Inconsistencies in his conception of progress may seem apparent to late-twentieth-century and twenty-first-century readers. Science and technology studies scholars in particular might find themselves yelling at Hank "Don't you understand that users matter? You can't just place inventions in a different time and place and re-create nineteenth-century American culture!" But Twain lacked the acuity of hindsight, and the novel treats the issue of invention-motivated progress seriously. Hank's reign, though short, is astoundingly productive. In addition to ending slavery, he promotes freedom of religion, eschewing "the Established Church" for a "go-as-you-please one." He even advocates "unlimited suffrage ... [for] men and women alike" (514) decades before women would be granted the vote in the United States. Throughout this episode, the Yankee sings the success of his experiment. He announces that he was able to improve social conditions for the masses in Camelot by re-producing nineteenth-century inventions. If Hank's emphasis on re-building nineteenth-century systems seemed amusingly self-serving in earlier episodes, this chapter vindicates his presumption that electrical and mechanical inventions could correlate to progressive social change.

Hank achieves this erstwhile success by prioritizing the public good over the public eye. And Twain accentuates his protagonist's atypically humanist orientation in this episode by bringing together Hank's private life as a man and his public life as the inventor of civilization.[38] In "Three Years Later," shortly after Hank unveils his utopia, readers learn simultaneously that he has married Sandy and fathered a daughter named "Hello-Central" after a telephone-operating hub. Seth Lerer (2003) convincingly argues that this birth links ideal nineteenth-century domesticity with nineteenth-century communication systems. However, this marriage of characters and ideas fails quickly. Shortly after the Yankee unveils his "civilization," he becomes so absorbed with his family life that he loses control of his modern empire. When his infant daughter falls ill, Hank leaves the spotlight he fought so hard to build and then to occupy. The power vacuum that ensues sparks two wars—one waged the old-fashioned way with swords and chain mail, the other with fantastic, electrified weaponry. Hank cannot maintain the illusion that his inventions are singularly progressive; their plural uses and connotations resist repression.

"WAR!"

The first war in the novel begins while Hank waits anxiously at his daughter's bedside. In chapter 42—simply titled "War!"—one of his newly trained journalists reports:

> Arthur had given order that if a sword was raised during the consultation over the proposed treaty with Mordred, sound the trumpet and fall on! for he had no confidence in Mordred. Mordred had given a similar order to *his* people. Well, by and by an adder bit a knight's heel; the knight forgot all about the order, and made a slash at the adder with his sword. Inside of half a minute those two prodigious hosts came together with a crash! (330)

According to this journalist's account, inane human error sparked this war. By staging an innocuous mistake (with biblical connotations) as the root cause of tens of thousands of deaths, *A Connecticut Yankee* suggests that social relations are intensely fragile because they are systemically interconnected: a small external perturbation can have catastrophic, unpredictable effects on the social system as a whole. In this case, one soldier's self-preservation instincts can set off a chain of violent events because of the unstable intricacies of a militarized social system. Ultimately, the misinterpreted sword precipitates a quick transference of power, which puts the Yankee and his trained apprentices in a defensive position. In this position, his dynamos take on more sinister social meanings than they had only a few chapters earlier.

The destructive potential of Hank's inventions becomes kinetic in the Battle of the Sand-belt. At the beginning of the novel, when Hank began to construct his electrical systems, he imagined himself at a metaphorical switch, turning on a light. At the novel's conclusion his power similarly resides in the buttons he figuratively and literally pushes. The parallel imagery in these scenes emphasizes the ostensible disinterestedness of artifacts themselves. The electric button ambivalently closes or breaks a circuit; it represents both connection and disconnection. Crucially, the button's appearance of neutrality is in itself political: the narrator's emphasis on the easy-to-operate switch highlights the fact that Hank can illuminate or kill with the same slight motion of his finger.[39] This scene registers how electrical devices can detach users from broader questions of cause and effect—an issue I explore in great detail in the next chapter.[40] With a quick re-tooling, Hank's supposedly civilizing systems easily become cataclysmic. The ease with which he completes his modifications hints that creative, destructive, and mundane potentialities simultaneously coexist with every interaction that is mediated through the Yankee's electrical inventions or the other systems they represent.

"The Battle of the Sand-belt" rhetorically resembles "Merlin's Tower" and "The Restoration of the Fountain" more than "War!" (the chapter it directly follows). Like previous showdowns, this episode pastiches detailed accounts of circuitry with more emotional accounts of their effects. It begins with Hank stationed in a bunker with his loyal assistant Clarence and a few nameless allies. In a lengthy conversation, he and Clarence discuss how to build a cost-effective, rationalized apparatus, and he ultimately decides: "You don't want any ground-connection except the one through the negative brush. The other end of every wire must be brought back into the cave and fastened independently, and *without* any ground-connection. Now, then, observe the economy of it. …You are using no power, you are spending no money, for there is only one ground-connection till those horses come against the wire" (541–542). This two-page passage offers a thorough description of the hardware of a simple and efficient electrocution apparatus of Hank's invention. His attention to detail in this passage is realist in multiple senses of the word: it conforms to the conventions of literary realism by setting a scene with mimetic detail, and it describes a system of execution that would technically work as long as the wire fences were not rushed by a horde of knights simultaneously.

As in "Merlin's Tower" and "The Restoration of the Fountain," Hank's tone changes from scientific to sublime as he shifts from describing cause to effect. Most of the Battle of the Sand-belt is told in cadences that express exhilaration rather than anxiety, as in the passage I took for an epigraph: "*There* was a groan you could *hear!*" This chapter is littered with such exclamations; Hank evidently finds electric currents equally thrilling whether they kill people or illuminate the sky. The Yankee's buoyancy in such passages echoes the shift into ecstatic narration in the earlier showdowns. This parallelism reveals how the Boss conflates violence with entertainment.

Further resembling the "Merlin's Tower" episode, Hank overemphasizes his bodily control over the electric current throughout the "Battle" chapter. Although he had described a circuit that would ground itself through intruders' bodies with no need for external intervention, Hank repeatedly manipulates the current, turning it on and off until he "shot the current through all the fences and struck the whole host dead in their tracks" (564). Such passages highlight both the raw power of electricity and the enthusiasm with which Hank controls it. Twain's descriptions of the method and the moment of destruction constitute another hybrid image: the apparatus Hank devises is both a deadly defense mechanism developed by an expert at building war machines and an amazing toy that demonstrates how easily electric power can make the savvy American almost invincible. Hank's conceptions of electricity as rational and sublime appear inextricable in this chapter.

Diverging from his accounts of earlier electric spectacles, however, Hank qualifies his excitement in this chapter with a note of plaintiveness. While most of the knights die anonymously en masse, confirming the dominance of Hank's electric war machine, the narrator describes one specific electrocution as "awful":

> He was near enough now for us to see him put out a hand, find an upper wire, then bend and step under it and over the lower one. Now he arrived at the first knight—and started slightly when he discovered him. He stood a moment—no doubt wondering why the other one didn't move on; then he said, in a low voice, "Why dreamest thou here, good Sir Mar—" then he laid his hand on the corpse's shoulder—and just uttered a little soft moan and sunk down dead. Killed by a dead man, you see—killed by a dead friend, in fact. There was something awful about it. (562–563)

Throughout the battle, Hank uses electricity to supplement his power over his enemy, but this moment captures the macabre fact that the electric current can invisibly transform a friend into a conductor. This scene renders relationship between electrical and human interconnection ominous, positioning an electrocuted corpse as the uncanny inversion of Hank's prosthetic, bodily use of electricity. The complementary images of Hank's electrified strength and the knight's radical vulnerability remind readers that the life spark and the death spark are equally functions of linking bodies and electricity.

The timbre of this chapter lends a new dimension to the plural characterization of electricity. For this isolated moment of accentuated pathos, Twain depicts electricity as more than a source of style, a scientific phenomenon, or a joke. It is also a force that affects and sometimes extinguishes human life. Still, the sympathy Hank feels for this particular electrocution doesn't imply that the "awfulness" of this death lies in the electrical system that caused it. Since the Battle of the Sand-belt appears after a long description of a nation-wide war waged the old-fashioned way, this haunting death illuminates the congruence between modern and pre-modern war: both transform friends into potential killers. Therefore, the mass of decomposing flesh of the 25,000 knights Hank kills by pressing a few electric buttons implies that new inventions can change the odds of a battle, and they can make it appear more spectacular, but they cannot change the essence of war or the fact that humans wage wars in the first place. These parallel instances of violence illustrate that the modern era is continuous with the past, not disjoint from it: the atavism of historic wars remains imprinted, palimpsest-like, on the modern battlefield.

In this context, it is important to note that the only temporal disruptions in the novel—Hank's moves back and forth in time—are effected by violence, not electrical (or other) inventions. Hank arrived in sixth-century Camelot after an employee knocked him on the head with a crowbar, and he returned to nineteenth-century America by way of Merlin's malevolent magic. Thus, the novel designates brutality as a powerful, consistent force in human existence. In so doing, *A Connecticut Yankee* hints that this tendency toward destructiveness remains imprinted on the inventions Hank correlates with civilization. The very artifacts that Hank and other determinists hope will enact change instead recapitulate existing power structures. Alternately awful and awe-inspiring, the Battle of the Sand-belt demonstrates both the enhanced power that electricity offers the modern American and the fact that electricity's uses are constrained by the values of the people who harness and use it.

CONFLICTED PARADIGMS OF PROGRESS

The dramatic shift in Hank's behavior from "Three Years Later" to "The Battle of the Sand-belt" echoes a historical shift in the American conceptualization of progress. Early-nineteenth-century republican thinkers understood scientific or mechanical innovation as a means to attain social progress, which they defined in terms of overcoming religious and social oppression. As Leo Marx notes (1994, 250), "The idea of history as a record of progress driven by the application of science-based knowledge was ... a figurative concept lodged at the center of what became, sometime after 1750, the dominant secular world-picture of Western culture."[41] Hank spends the entire novel claiming to champion similar ideals. He frames his exploits most explicitly in terms of this early conception of progress in his "Beginnings of Civilization" chapter, in which he describes the various devices and systems he created in terms of "human liberty" and "human thought." He explains that laborers "had formerly been worked as savages" until he makes "pretty handsome progress" (118).

In "Three Years Later," Hank faithfully returns to this conventional progressive goal. When Hank situates electricity as one useful tool among many, rather than as a symbol of his own prowess over supposedly primitive Others, he genuinely promotes social change. But when he opts to defend his inventions over human life, he dramatizes the emergent belief that inventions could constitute an end in themselves, rather than a means to attain human equality. *A Connecticut Yankee* approaches these contested constructions of progress by transitioning from a linear narrative into a circular one. Before the Battle of the Sand-belt, the Yankee seemed to reenact the popular history of American electrical development. He emulated Franklin's invention of the lightning rod, he invented a simple battery-operated button, and finally he (implicitly) constructed a light and power system. The Battle of the Sand-belt interrupts Hank's linear progress and his growth as a character, knocking him back to the future, to the nineteenth-century setting in which the novel began. This disruption of Hank's linear narrative of progress begs the question: What went wrong? More specifically, what activated Hank's shift from one paradigm of progress to the other?

Most scholarship about technology in *A Connecticut Yankee* has demonstrated that the exploitative or destructive potential of Hank's regime remains present along the margins of his short-lived utopia. Hank's modernized Camelot is conspicuously monolithic; he has commandeered the nation's systems under authoritarian supervision while claiming to democratize them.[42] As Gregory Pfitzer (1994) and Richard Slotkin (1994) have recognized, Hank justifies his domination of Camelot (at least in part) through analogy to westward expansion; after he makes his inventions public and purportedly enlightens the "white Indians" of Camelot, he gets "ready to send out an expedition to discover America," implicitly preparing to reenact the violent colonization of the "new world" by franchising his model of modern society.

The instability of this utopian conclusion also comports with the literary realists' resistance to happy endings. Though *A Connecticut Yankee* scarcely could be identified as "realistic" in the colloquial sense of the word, it shares with literary realism the aim of satirizing the revival of romance narratives in American culture. Prominent figures who promoted these literary values derided the overuse of convenient resolutions in literature, classifying such conventions as harmful to readers.[43] Dovetailing with the works of contemporary literary realists, including those by his good friend William Dean Howells, Twain chooses to destroy this domestic resolution and to replace it with what Wai Chi Dimock calls an "economy of pain" (1991, 67–90). In fact, the domestic resolution can even be said to give Hank the false sense of stability and closure that ultimately result in his downfall.

In this context, the novel hints that familial life and individualistic desires might be at odds with Hank's increasingly systematized version of the modern world. Since Twain wrote this book at a moment when large-scale systems (including railroads and power-distribution systems) were being mismanaged by monopolies, his conflicted account implies anxiety about how easily these complex systems could be corrupted. Hank explicitly voices such concern early in the novel, when he recognizes that the sixth century can offer opportunities for him that the dawning twentieth century cannot: "Look at the opportunities here for a man of knowledge, brains, pluck and enterprise to sail in and grow up with the country. The grandest field that ever was; and all my own … whereas, what would I amount to in the twentieth century? I should be a foreman of a factory, that is about all; and could drag a seine downstreet any day and catch a hundred better men than myself" (52). Although most critics have read Hank as a confident inventor, this passage suggests that Twain's spokesperson for modern progress is already intimidated by it.

Specifically, Hank craves the opportunity to be somebody—to be an important individual rather than one node in a complex network.[44] Consider how he relishes earning the title "The Boss" in the same chapter in which he expresses the above concerns: "There were very few THE's, and I was one of them. If you spoke of the duke, or the earl, or the bishop, how could anybody tell which one you meant? But if you spoke of The King or The Queen or The Boss, it was different" (103). Hank clearly yearns to stand out as an exceptional individual and believes that he could not rise to such prominence in the dawning twentieth century. Still, he would rather rebuild modern systems than withdraw into a pastoral fantasy of romantic individualism or domesticity. Hank's desire to re-create every system he can imagine—including educational, economic, electrical, communication, and transportation systems—illustrates the allure of "systems thinking" to this character. Although he has the capacity to build himself a generator and to enjoy modern conveniences on his own or with his family, he is torn between his personal inclinations and his belief that modern life must be complex and networked.

Ultimately, the brevity of Hank's utopian endeavor emerges from the tension among his three underlying desires—his need to feel significant as an individual, his emotional connection to his family, and his proclivity for systems thinking. This plurality might indicate why the electric power system became this character's organizing metaphor: the image of Hank standing with his "hand on the cock, so to speak" (120) represents the importance of Hank's decisions as a single actor on the system writ large. Indeed, his inability to choose between individualistic and systemic paradigms might also explain why he interposes his own body in descriptions of electrical circuits that can operate autonomously, such as the lightning rod or the electric fence: he wants to understand himself as an individual whose very being is important and as a crucial part of a complex social and technical system.

Twain exacerbates the incompatibility between Hank's role as an individual and as a part of a system in the novel's nested conclusions. After the Battle of the Sand-belt, the narrator weaves his distinctive electrical imagery together with other thematic threads, effectively illuminating the consequences of Hank's pushing of electric buttons. At the close of the Battle, Hank's masterful use of electrified weapons renders him victorious but also vulnerable. Despite his ostensible success, Hank cannot reach his family because he destroyed his own telegraph and telephone systems early in the battle. Isolated, he begins composing letters to Sandy, which (combined with his edited diary) transform into his personal memoir, and, in turn, into the substance of the novel. Thus, the conclusion reveals that *A Connecticut Yankee* represents, in part, a failed communication between husband and wife.

From this point on, Clarence and Mark Twain (the frame narrator) take over Hank's personal memoir. Clarence's postscript explains that, by "winning" the war, Hank inadvertently barricaded himself and his men in a cave surrounded by dead, rotting flesh. The Yankee's inventions could defend against the knights' swords but not against their contagion. As Clarence explains, "If we stayed where we were, our dead would kill us; if we moved out of our defences, we should no longer be invincible. We had conquered; in turn we were conquered."[45]

In this way, after the Battle of the Sand-belt, Hank appears as both a disconnected individual who cannot communicate with the outside world and a hyper-connected organism helplessly in touch with death and disease. The novel's earlier distinctions between communications and power—between figuratively and literally annihilating space—break down. Hank's perceived ability to manage systemic interconnections also collapses. After Hank is weakened by a wound and by his proximity to the rotting corpses, Merlin sneaks into the rebels' bunker and takes revenge. The magician enchants the Yankee to sleep for thirteen centuries, removing Hank in time and space from his family and his circle of temporarily victorious allies. The novel ends with Hank back in the nineteenth century, crying, "Hello-Central." He dies as he calls for his sixth-century daughter and for emotional and electrical connections of the sort he had developed and destroyed in Camelot. Metaphorically stripped of all connections, the novel concludes with an image of Hank's final failure: "He was getting up his last 'effect;' but he never finished it" (575). Hank dies a dynamo without a system—a symbol of wasted potential.[46]

THE EXAMINATION OF A HAZARDOUS CONCEPT

By reading Hank as a "dynamo" and emphasizing the uniqueness of electricity in *A Connecticut Yankee*, I do not aim to dismiss the relevance of other prominent themes. In fact, in the final three chapters of the novel Hank's manipulation of electrical systems blurs together with other themes, including communication and violence and imperialism. This polysyndeton matters. The infamous incoherence of the novel and the narrator's proclivity for hybrid analogies attest to the fact that this character's root desires are plural: he wants individual power and the modern conveniences of complex systems, but he cannot reconcile these desires with his familial and democratic ideals. By projecting the term *technology* onto this novel, present-day readers fallaciously collapse plural ideas into a singular concept. In so doing, they misconstrue this novel as a reaction to technological changes, rather than a playful exploration of the metaphorical and actual possibilities for individuals, energies, artifacts, and systems.

The technological fallacy has delimited the types of questions readers have asked about this novel and other cultural artifacts from this era. It has inspired critics to finish *A Connecticut Yankee* wondering why Merlin "wins," what his victory implies about the contest between technology and magic, and whether Hank's character was meant to satirize or celebrate American ingenuity. These questions are reductive; they seek singular answers that inevitably fail to appreciate that the richness of this novel lies in its multifariousness. If we set aside questions of technology and ask how Twain engages with the theme of electrification in this novel, we get a range of answers (or at least new questions) about the slipperiness of metaphors and reality—and about the incommensurable yet coexisting desires that define human life. If we expand our range of inquiry and ask what electrification meant to Samuel Clemens, the man who "played" the author Mark Twain, we find a similar host of answers: it was impressive to behold, it might be the best metaphor for describing the thrill of good showmanship, its medical applications might save the life of his ailing wife. If we expand our focus still further, we may ask what plurality and ambiguity, individualism and systems, mean in this text and in this author's larger body of work. At this scale, we can better appreciate the emotional appeal of Hank's dashed hopes. Hank's inability to accept his own multitudinousness—to reconcile his proclivity for technics with his desires for love, recognition, and power—pokes fun at American pragmatism, but it is also aestheticizes this condition of cognitive dissonance as distinctly and beautifully human.

My reappraisal of critics' use of the word *technology* doesn't subvert the significant work that grew out of this scholarly tradition. Instead, it urges readers to recognize the term's plurality and contingency—and to question why scholars who were acutely attuned to the subtleties of language would project an anachronistic concept onto their readings of this novel. The great Twain scholar Henry Nash Smith begins his 1964 book *Mark Twain's Fable of Progress* by discussing the invention of the terms *individualism* and *capitalism*, but he unreflectively incorporates the word *technology* into his understanding of the novel and of the historical moment in which it emerged.[47] As Leo Marx meditated (1997, 967), "It is curious that many humanist scholars ... have so casually projected the idea back into the past, and into cultures, in which it was unknown"—especially curious, I would add, in light of the attention scholars have paid to other emergent terms.

If *technology* has worked its way into even the most observant literary criticism, that is because the term arose from many of the cultural tensions that *A Connecticut Yankee* addresses, including the emergence of new artifacts and the changing conception of history itself as a record of incremental improvements (or progress). The lack of attention given to the word *technology* doesn't reveal carelessness; it exposes the paucity of critical terms available to describe the interplay among people, systems, discourses, practices, and artifacts (old and newly invented). Thus, the stakes of redressing the technological fallacy exceed our reading of one novel. By acknowledging that our shorthand term *technology* can obscure more than it elucidates, we might appreciate more fully the difficulty Twain confronted in composing a novel that teased out (and just plain teased) the relationship between political ideologies and material culture. And we might reconsider the careless use of such terms, even today, when we try to describe the place of human life in a world that may continue to feel circumscribed by complex, macro-scale systems.

2 SHOCK AND SENSIBILITY: THE RHETORICS OF ELECTRIC EXECUTION

To grasp the broader stakes of the technological fallacy, this chapter considers its relevance to the history of capital punishment. In chapter 1, I identified this fallacy in the twentieth-century scholarship that projected what Nye (2007, 15) calls the "unusually slippery term" *technology* onto *A Connecticut Yankee*. Yet redressing this fallacy entails more than avoiding anachronism. It involves recognizing how the American fascination with apparatuses can inflect the stories we tell about our past, our present, and our future. Just as a laboratory apparatus will determine the types of answers any experiment will yield, a cultural fascination with *technology* in general—or with a specific technical artifact like electrical power—can shape our narratives about and concomitant understandings of light and life and death.

Historians of capital punishment have been critical of this very tendency. For example, the professor of jurisprudence and political science Austin Sarat argues as follows (2001, 64): "Technology mediates between the state and death by masking physical pain and allowing citizens to imagine that state killing is painless. The language of law works hand in hand with this technology to veil the ugly realities of execution, separating cause and effect, and making it unclear who is actually ordering and doing the killing." *Technology*, in this context, is more than the abstract, singular synonym for *tool*; it is a technique of obfuscation.

Sarat persuasively argues that modifications to the practice of execution shifted control into the hands of experts, privatizing the process. He, Stuart Banner (2002), and other scholars have demonstrated that inventions such as the "upright jerker" (a modified apparatus for hanging), electric execution, and lethal injection served as incremental steps toward routinizing and rationalizing death. These scholars have revealed that the promise of a technically perfectible execution method has had the power to displace questions of whether modern American states should practice capital punishment with questions of how these governing bodies could practice it most humanely. Those who think of state execution in these terms would be more likely to surmise from botched executions that the technology, rather than the practice of capital punishment, is imperfect.

The mechanism by which the public dialogue shifted from questions of whether to questions of how demands careful scrutiny. This shift in thinking about execution as a technical rather than an ethical problem demonstrates the reductive, technologically fallacious thinking that this book draws into question. This paradigm shift coincided with the implementation of electric execution, and we can better understand the change and its reverberations in our own time by exploring how historical figures explained this phenomenon in their own terms.

Before the Civil War, the debate about whether modern states should execute capital sentences was kept in the public eye by a consolidated, energetic movement to abolish the death penalty. Activists committed to this cause contended that no civilized society should condone a practice as gruesome as hanging.[1] The campaign lost momentum during the 1850s and the 1860s, as abolitionists shifted their attention from the brutality of execution to that of slavery and war. A wave of apathy followed. The historian of capital punishment Roger E. Schwed (1983, 15) has rationalized this trend by arguing that "the severe cruelties of the war rendered insignificant for many the executions of persons who were, after all, convicted murderers."

Indifference to incarcerated people was one reason why many condoned the adoption of the electric chair. The symbolic power of electricity was another. In fact, before the state of New York signed the first Electric Execution Act into law, many Americans responded with relative indifference to the loss of innocent lives as a result of incidental electric shock. Coverage of electrical accidents attended more closely to the energy source than to the loss of life, suggesting that the "death spark" enticed many to wonder how it worked in tandem with or in the place of other ethical questions they might have regarding the development of potentially fatal power systems.

The *Buffalo Commercial Advertiser*'s description of the first known death by electricity (reprinted in the *New York Times* on August 11, 1881) exemplifies this tendency.[2] The article, headlined "Killed By an Electric Shock," offers a detailed account of the electrical apparatus involved in the incident:

> One of the most peculiar accidents we have been called upon to report in a long time occurred last night at the station of the Brush Electric Light Company on Ganson-street. A few feet from the door is the generator [which] is constructed of iron and copper wire, and weighs about 4,800 pounds. The armature makes about 700 revolutions a minute, and generates a current equal to 225,000 cells of a battery. On the end of the machine is a brass and copper cylinder, termed a commutator, where all the current generated accumulates. … The curiosity of many visitors could not be satisfied until they had experienced a shock. This is received by several persons joining hands, thereby making a circuit, those on the end touching one of the "brushes," when a tingling sensation is felt. There is no danger in this.
>
> Among the visitors last evening was George L. Smith … . He appeared about 10 o'clock and began to examine the apparatus. [After being evicted and returning later in the evening, Smith] leaned over the railing in another attempt to reach the copper. He seized two strips, one in each hand. Instantly a circuit was formed, and he dropped onto the railing rigid. … It was seen at once that Smith was dead.

This report foregrounds public interest in the electrical plant and subordinates its discussion of the man who died there. Even if its emphasis on the "curiosity of many visitors" seems morbid by twenty-first-century standards, this article piques readers' curiosity still further by raising more questions than it answers. It slips into the passive voice in the middle of the sentence that recounts the accident. Stating simply, "a circuit was formed," it obscures the interaction the article purports to address. The reader is left to wonder about how this circuit felt to Smith in his last moments. The moment of actual electric shock remains unfathomable, described only with the temporal modifiers "instantly" and "at once." The piece also offers mechanical minutiae in the place of other salient information. A report of the armature's revolutions per minute cannot elucidate why so many visitors would yearn to feel the generator's "shock" for themselves. Nor can these details explain why Smith would return to the electrical generator a second time—especially if the physical sensation conferred was simply a "tingling," as the column suggests. The litany of technical specifications seems to demystify electricity by making the energy appear quantifiable, but it cultivates an aura of mystery by failing to address the most striking aspects of the accident.

Asymmetrical reports similar to the one quoted above proliferated during the late 1880s, registering and reinforcing the public's deep fascination with electrical power. Recorded responses to electric death corroborate this impression. News of Smith's fatal electric shock famously inspired Alfred Southwick to dream up the electric chair: he apprehended this accidental death as an interesting idea, and his reaction was not anomalous. After the first state-sanctioned electric execution, sales of home electro-medical devices increased markedly—a correlation which suggests that many responded to the news of electric execution with intrigue rather than outrage.[3]

The impulse to emphasize the instantaneousness of electrical transmission over the tragic loss of life was a significant consideration in the selection of a new method of execution. It meant that by 1886, when the state of New York appointed a commission "to investigate and report at an early date the most humane and practical method known to modern science of carrying into effect the sentence of death in capital cases," there had already been a precedent for understanding electric death in quantifiable terms (Gerry et al. 1888, 3). And that was not the only factor that contributed to electricity's appeal. Proponents of the death penalty required a solution that was mysterious as well as "humane and practical," since capital sentences could only be imagined to deter crime if they could intimidate potential criminals. As I will demonstrate below, early advocates of electric execution suggested that electricity was uniquely suited to this paradoxical task. To legislatures concerned with the public perception of capital punishment, electricity was an attractive alternative to hanging for interrelated socioeconomic, technical, and representational reasons, although the former has been overly pronounced in most studies.

Histories of capital punishment typically emphasize the corporate motivations that promoted the adoption of electric execution. They often highlight Thomas Edison's endorsement of this new form of punishment and his related hope that electric execution would damage the sales of his competitor, George Westinghouse.[4] Known then and now as "The Battle of the Currents" or "The Battle of the Systems," this famous rivalry began when Thomas Edison—the "father" of direct-current power generation—advocated the use of alternating current in the death chamber in order to frighten customers away from the system George Westinghouse had patented. AC power does affect the human body differently than DC. People who come in contact with a live wire from an AC generator freeze. In contrast, people who touch a live DC wire can drop the wire relatively unscathed. Nonetheless, Edison's motivation for advertising this distinction was transparently competitive; he suggested that Americans call the process of execution "westinghousing" (Brandon 1999, 9).[5] Westinghouse fought back, hiring lawyers to challenge the constitutionality of electric execution and even refusing to sell New York (and later Ohio) prisons AC generators.

The electric chair ultimately was powered by a Westinghouse generator, but that fact didn't have the effect Edison hoped for or the effect Westinghouse feared. Instead of eliciting fear of the power of AC current, newspaper coverage of the Battle of the Currents incited public curiosity. Debates over which entrepreneur-celebrity would "win" the battle helped shift the debate from whether to how by overshadowing discussion about the actual fates of people condemned to death. For example, on August 7, 1890, after William Kemmler became the first man to be executed by electricity, the *New York Times* ran two articles about his death: a front-page piece about how terribly the execution was bungled ("Far Worse Than Hanging") and a second-page column that hypothesized "Westinghouse Is Satisfied. He Thinks There Will Be No More Electrical Executions." The latter article undermines the former by implying that the abolition of the electric chair would serve Westinghouse's, and not society's, best interests.

Although the Battle of the Currents remains fascinating to history buffs today, this rivalry was not the most salient consideration involved in the adoption of electric execution. While Edison, Westinghouse, and readers who were interested in their feud fretted about the type of electrical current that would be adopted, lawmakers were more interested in changing the way this new form of execution would be depicted by the press. They explicitly wrote their concerns about representation into the law itself.[6] The Electric Execution Act didn't define the parameters by which state prisons could acquire apparatuses for electric executions. It didn't explain how electricity would be directed through a human body, but it did include guidelines about how these executions could be described to the public: "No account of the details of any such execution, beyond the statement of the fact that such convict was on the day in question duly executed according to the law at the prison, shall be published in any newspaper. Any person who shall violate or omit to comply with any provision of this section shall be guilty of a misdemeanor." The law's insistence that reports of executions should be reduced to a "statement of fact" suggests that lawmakers wanted state-sanctioned electric execution to be approached in the manner of Smith's death: in a technical idiom with an emphasis on the numbers.

The gag clause was not in fact necessary to produce this desired effect. Even after its repeal in 1892, electric executions were reported in a similar manner. Both before and after the repeal of this clause, the use of electricity to execute capital sentences functioned like a magnetic field, attracting attention to the power source itself. Advocates of the new process concentrated on the apparatus to maintain that electricity painlessly and immediately extinguished life. Critics focused on the electric charge to ridicule the notion that an energy so thoroughly associated with vitality could kill effectively or at all.[7] Regardless of their ethical stance on the practice, most narratives about electric execution from this era emphasized the components that we would call technological today: the electrical charge, the power generator that supplied the current, and the chair that delivered the shock to a human body. Specifically, they focused with rapt attention on the moment that electricity passed through the body.

The decision to spotlight this single instant was not obvious. It involved marginalizing the social, legal, and political systems that enabled the "death spark" to function. As a rare dissenter, Dr. Shandy, observed shortly after viewing firsthand the first execution by electricity: "The preliminaries of electro-thanasia [a term that predated the word *electrocution* in the American vernacular] are far from pleasant to contemplate. Alongside of those for hanging they are pretentiously horrible. There is something more than weird in the preparation of the machine, the deliberate fixation of the victim, the adjustment of the electrode, the thousand deaths in contemplating one; which more than offset the quick though damnable taking off. The horrors, though hidden, are nevertheless felt. There is something else to be thought of than the mere quickness of death."[8] The fact that other reporters would habitually elide all of these details warrants scholarly consideration.

In this context, avoiding the technological fallacy becomes even more challenging and worthwhile. If we, too, focus on the adjective (*electric*) rather than the noun it modifies (*execution*), we naturalize the choice to read this process metonymically in terms of its apparatus. In other words, the technological fallacy can encourage readers to accept unquestioningly the emphasis placed on the device and to ignore the social, economic, and legal factors that sanctioned electric execution. By organizing our history in this manner, we risk understanding the invention of electric execution as a casualty in the Battle of the Currents, rather than as a policy that helped Americans continue to rationalize the death penalty long after many other industrialized nations abolished the practice.

To open this history up to interesting and underexamined lines of interrogation, I here apply the approach of chapter 1 to different ends. That is, I explore the entangled histories of literature and electricity as a particularly rich starting point for interrogating the changing social meanings of capital punishment. I begin this chapter by exploring critics' and advocates' overemphasis on instantaneity and the electric current. I argue that writers' fixation on these aspects signaled and helped to initiate a transformation in the way most Americans understood capital punishment. I then turn to two successful but anomalous American novels that were based on actual executions: Gertrude Atherton's once-popular but subsequently forgotten 1897 work *Patience Sparhawk and Her Times*, which re-imagined Carlyle Harris's 1892 trial and 1893 execution, and Theodore Dreiser's 1925 masterpiece *An American Tragedy*, which revisited Chester Gillette's 1906 trial and 1908 execution. Since their subject matter was originally represented with the narrative strategies I discuss in the first half of this chapter, both novels can be said to model alternative ways that these cases might be understood.[9] By challenging the conventional narrative strategies used to represent this practice, both novels illuminate social and material dimensions of the execution process that are obscured in journalistic accounts. In this sense, their fictions accord with the material turn in literary studies and in science and technology studies: they make conspicuous the many decisions that coalesce into practices and apparatuses.

By turning to fiction at the midpoint of this chapter, I do not aim to offer a comprehensive close reading of either work. Though my readings contribute to ongoing conversations about the legacy of each author, my focus is limited: the electric chair scenes on which I focus make up only 112 of their combined 1,362 pages. But the relatively small role the chair plays in these lengthy novels is precisely the point. By situating execution in broader, systemic terms, these novels remind readers that the electric chair doesn't operate in a laboratory setting under ideal conditions. The circumstances that place a human life in the chair are much more complex than most journalistic, technical, or legal accounts would have readers believe.

Dreiser and Atherton demand attention—in their own times and today—because they provoke readers to stop perceiving the electric chair as an isolated invention and to begin considering the socioeconomic and cultural systems that give this artifact meaning. Crucially, readers seemed to enjoy the challenge. Both works were successful enough to remain in print for generations.[10] In fact, *An American Tragedy* has never fallen out of print since the time of its publication. By reading these fictional counter-narratives alongside representative reports, I illuminate the historical tension between the questions of whether and how American states should execute criminals—a debate that remains urgent and unsettled today as multiple states continue to practice "technologically" advanced, seemingly humane forms of capital punishment and shortly after Tennessee considered a return to electric execution in 2014.

EXAGGERATING INSTANTANEITY

The conceptual shift from whether to how was nonlinear. Americans were fascinated with how electricity could kill before they seriously entertained the question of whether it should be used to do so. They understood the fatal qualities of this energy long before they coined the term *electrocution* or designed the first electric chair.[11] Benjamin Franklin's *Experiments and Observations on Electricity* (originally published in 1751) discussed the electric extermination of animals. It also imagined employing electricity to punish those who would threaten the state. In a thought experiment called "The Conspirators," Franklin described how a gilt crown on a painting of the King could be electrified to punish those who would damage his likeness: "If now the picture be moderately electrified, and another person take hold of the frame with one hand, so that his fingers touch its inside gilding, and with the other hand endeavor to take off the crown, he will receive a terrible blow, and fail in the attempt. If the picture were highly charged, the consequence might perhaps be as fatal as that of high treason" (Franklin 1751). Here, in a scientific treatise about electricity, Franklin focused more on the nearly magical power of the electrical charge than on ethical implications of executions for acts of treason. He was ahead of his time in more ways than one.

The vision of Franklin's experiment would not be elaborated upon for another hundred years, and it would never be realized in the form he imagined. In his experiment, "conspirators" unwittingly shock themselves, much like the knights who happen upon the Yankee's electric fence during Twain's "Battle of the Sand-belt." In contrast, victims of state-sanctioned electrocution wait in prison, aware that they will eventually die by the uncanny shock of electricity. Nonetheless, the psychological strain of anticipation remained understated in most *fin de siècle* narratives of electrocution. The presumption that electric death would be instantaneous and easy to orchestrate suffused narratives about this method of capital punishment, both before and after it was imperfectly put into practice.

After Franklin, the earliest advocates of state-sanctioned electric execution that I could discover were staff writers for the popular science monthly *Scientific American*. The magazine first proposed the use of electricity for capital punishment in a January 8, 1876 article titled "Electricity as an Executioner." Published ten years before the state of New York assembled a commission to investigate an execution method to replace hanging, the piece offered several disparate lines of reasoning to build its case. Its kaleidoscopic account is worth analyzing at length because it models the varied arguments proponents would use to justify the electric chair for years to come. In fact, many of its strategies resonate with *A Connecticut Yankee*, registering the diffusion of these ideas before the first electric execution took place in 1890.

The first claim the article submits is representational: electricity could spare spectators from seeing or reading about the "revolting scenes" of hanging. Its second claim is humanitarian: extrapolating from Hermann von Helmholtz's experiments with nerves, "Electricity as an Executioner" surmises that the electric current would kill before the victim could experience pain. And the abatement of the victim's pain and the spectator's discomfort in observing that pain were not the only purported benefits. The article also implies that electricity is the only agent ambiguous enough to enable capital punishment to appear both humane and disturbing enough to deter future crime: "The same ignominy which attaches to the gallows would be transferred to this mode of destruction, while the peculiar death by lightning, which, among the ignorant of all nations and ages, has been the subject of profound superstition, would, without doubt, through its very incomprehensibility and mystery, imbue the uneducated masses with a deeper horror." Anticipating *A Connecticut Yankee*, this passage invokes the plural connotations of "death by lightning" to address the multiple roles electricity could play simultaneously.[12] In so doing, it illuminates the presumed social positions of the executed and the observer. Imagining potential criminals to be of the "uneducated masses," it supposes that this demographic would fear electric execution for its associations with divine retribution. Concomitantly, the article implies that the educated spectator could trust electrical experts to generate a charge as swift and powerful as the unmediated lightning strike itself. This column frames this expected disparity between the executed person and the spectator as a selling point; as I will discuss below, Dreiser reframes this issue as a symptom of systemic classism.

If the varied assertions above were not convincing enough, "Electricity as an Executioner" adds a brief economic justification. It nonchalantly mentions that the "instruments would rarely need replacing and would last indefinitely for other executions"—as if the cost of rope and gallows was a deterrent for hanging. This rationale recalls Twain's "Sand-belt" scene in which the Connecticut Yankee explains how to build an electric weapon frugally. When Hank details his dynamo design, his interest in saving money appears ludicrous. As the sole owner of the electrical system, he would have no financial motivation to optimize expense (aside from a cultural logic that abhors waste, a subject I address in chapter 3). The absurdity of this conversation in *A Connecticut Yankee* and its earnest parallel in *Scientific American* reveal the influence of utilitarianism and capitalism on the cultural understanding of electrical systems. Both texts imagine that electrical systems can be understood as optimal (and therefore as "progress" from previous inventions) based solely on the metric of cost, regardless of whether those systems illuminate or electrocute.[13]

While "Electricity as an Executioner" presupposes the relevance of this economic argument, it offers an explicit defense of its scientific rationale. After its quick mention of expenditure, the article adds: "The teachings of Science are heeded and sought for in the building of prisons, in the management and care of convicts, and in every modern correctional system; and yet in so simple and easy a process as the extinguishing of human life they are utterly ignored." The contrapositive logic of this passage characterizes the dismissal of a scientific perspective as irrational. By insisting upon the pertinence of (singular, reified) Science in such a convoluted manner, this passage reveals the novelty of approaching the death penalty from a technical rather than a legal or ethical vantage point. By describing the "extinguishing of human life" as "simple and easy," the piece begins to reframe capital punishment as a question of how rather than of whether to extinguish a human life.

Scientific American proudly reiterated this argument nine years later in a piece titled "Electricity for Executing Criminals." The perspective proposed in 1876 and taken for granted by the same magazine in 1885 emphasizes technical simplicity over ethical complexity. Indeed, the latter article claims that years of electrical development had only served to render electric execution easier to enact than previously imagined: "How simple a process it would be to connect the place of execution in the Tombs with the system of electrical street illumination, so that electricity could be made the executioner of murderers!" This passage pre-figures Hank Morgan's example in two ways. It emphasizes the facility of repurposing light and heat systems to deadlier tasks, and it reduces the issue of taking life to the deceptively simple task of pushing a button. The article projects that "it would be possible to furnish the death seat with an automatic attachment so that the execution could be effected at a given moment by the action of a clock-like apparatus, and without the least involvement of the hand of the officer charged with the infliction of the death penalty." The passive conditional voice of this passage serves a similar function to its proposed "clock-like apparatus": it enhances the perceived displacement between cause and effect, obscuring the executioner's role in the manner that Austin Sarat described in the passage quoted above.

"Electricity for Executing Criminals" closes with an unattributed quotation: "What more scientific method than the one here proposed can be devised? Death would be instantaneous and perfectly painless, *while at the same time* the awfulness of the penalty thus inflicted would be profoundly impressive" (*Scientific American* 1885, 101, emphasis added). Reflections such as this one suggest that electric execution seemed ideal because it was fundamentally plural. Ultimately, "Electricity for Executing Criminals" employs an array of figurative techniques similar to those used in "Electricity as an Executioner" and in *A Connecticut Yankee in King Arthur's Court* in order to depict electric death as instantaneous, predictably controllable by scientific principles, and mysterious enough to frighten potential criminals.

Critics of the death penalty provocatively shared this article's understanding of electric execution as "scientific," "instantaneous," and "profoundly impressive." For example, William Dean Howells's 1888 essay in *Harper's Weekly* glibly accepts the technical simplicity of electric execution: "I understand that the death-spark can be applied with a *minimum* of official intervention, and without even arousing the victim. ... One journal has drawn an interesting picture of the *simple process*, and I have fancied the executions throughout the State taking place from the Governor's office, where his private secretary, or the Governor himself, might *touch a little annunciator button* and dismiss a murderer to the presence of his Maker with *the lightest pressure of the finger*" (Howells 1888, 23, emphasis added).[14] Echoing *Scientific American*, Howells emphasizes effortlessness with a smattering of diminutive adjectives: the process is "simple," the button that actuates it is "little," and activation requires only "the lightest pressure." However, he problematizes the presumption that the value of this practice should be measured in terms of the executioner's exertion.[15] After bringing to the forefront the ethical concerns that *Scientific American* had displaced, he sardonically suggests that Americans who support the death penalty ought to share the executioner's responsibility, either by allowing children to push the button or by adopting a lottery system as they do in the case of jury duty. His article raises a question: If the task were as "simple" as pressing a button, why would most Americans refuse to take on the responsibility?

Howells frames the practice of electric execution as a consequence of cognitive dissonance. He implies that Americans condone or ignore the practice because they don't have to take responsibility for it. And he pokes fun at the idea that they can read this practice as "progressive" simply because other applications of electrical power are considered "advancements": "The weight of learned testimony seems to be in favor of electricity, and there is apparently no good reason why this mysterious agent, which now unites the whole civilized world by nerves of keen intelligence, which illuminates every enterprising city, which already propels trains of cars and promises to heat them, which has added to life in apparently inexhaustible variety, should not also be employed to take it away" (Howells 1888, 23). Throughout this article (and other essays on the same subject), Howells insists that electricity is not *inherently* progressive or democratic; applying electricity to capital punishment doesn't render the practice any less barbaric.

Another journalist and fiction writer, Stephen Crane, began contributing to this conversation eight years later. Whereas Howells had derided Americans who hypocritically complied with the new Electric Execution Act, Crane tries to explain why they might countenance the law unreflectively. In an illustrated full-page article in the sensational New York weekly *The World* titled "The Devil's Acre,"[16] Crane suggests that Americans ignore the issue of electric execution because they lack an appropriate framework for understanding it. Like Howells, Crane never questions the idea that electric death would be instantaneous. He doesn't editorialize the groundskeeper's claim that the entire process of execution "takes about a minute" (Crane 1896, 23). Rather than emphasizing or debating the celerity of the alluring death spark, he takes interest in the setting of the execution chamber: the "low gray building," the scent of varnish, the single chair, the "wire almost as thick as a cigar." He lingers on these quotidian details in an attempt to make the electric chair real to his readers. He wants readers to imagine walking into a room and seeing the material apparatus. He believes this visage would challenge the idea of an instantaneous, nearly mythic "death spark." Crane makes this intention explicit when he shifts his focus from the chair to a discussion about the way Americans apprehend it.[17] In an uncommonly philosophical tone for the sensational paper in which this article appeared, he argues:

> We, as a new people, are likely to conclude that our mechanical perfection, our structural precision, is certain to destroy all quality of sentiment in our devices, and so we prefer to grope in the past when people are not supposed to have had any structural precision. As the terrible, the beautiful, the ghastly,

pass continually before our eyes we merely remark that they do not seem to be correct in romantic detail.

But an odor of oiled woods, a keeper's tranquil, unemotional voice, a broom stood in the corner near the door, a blue sky and a bit of moving green tree at a window so small that it might have been made by a canister shot—all these ordinary things contribute with subtle meanings to the horror of this comfortable chair, this commonplace bit of furniture that waits in silence and loneliness, and waits and waits and waits. (Crane 1896, 23)

Crane's description invokes the aesthetics of the sublime with which he apprehends the electric chair. But, he argues, most Americans fail to perceive the chair with awe. In fact, he suggests that most fail to perceive it at all. The rhythmic cadence of his phrase, "the terrible, the beautiful, the ghastly," elicits an impression of a steady march passing an oblivious populace by. Crane claims that sentimental ideals make Americans passive: the expectation of "romantic detail" impedes them from confronting the stakes of their own inventions. This article anticipates the tension between advancements in scientific capability and genteel cultural norms that he would explore in his masterful story, "The Monster," two years later.[18] In both his fiction and his newspaper copy, Crane hints that the tension between new inventions and residual social mores might be mitigated somewhat by more mimetic forms of representation. He doesn't explicitly call for change to the practice of electric execution (or, in "The Monster," of artificially extending life), but he demands a change in the way Americans comprehend these new capabilities.[19]

Crane's derision of sentiment and his emphasis on empirical description may seem to mimic the scientific perspective advocated by *Scientific American* in "Electricity as an Executioner," but Crane finds complexity where *Scientific American* saw simplicity. He contends that the electric chair cannot be understood in isolation. In a groping, convoluted way, he implies that accounts of the chair as an instantaneous lightning spark harmfully obscure the truth. By situating the chair in its physical context, Crane proposes an alternative paradigm for representing the apparatus. He challenges the fantasy of instantaneity by reminding his readers that the chair exists in space, portentously "waiting" before it is ever put into use. By suggesting that the waiting must be as meaningful as the execution itself, Crane insists that Americans should be made to recognize the

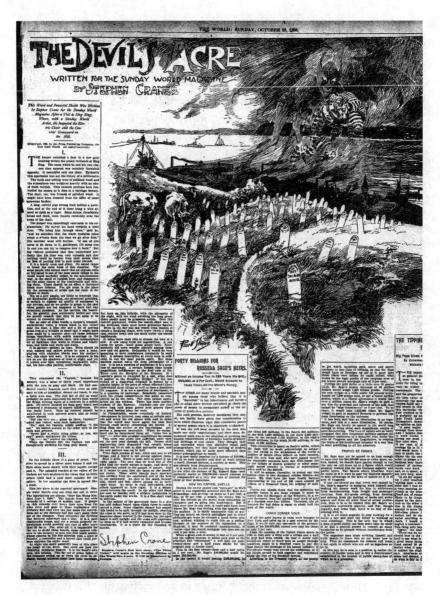

Figure 2.1
The first page of Stephen Crane's article "The Devil's Acre."

electric chair as a real apparatus in a real room and not as a fantastic agent of instant death.

This ethic of realism exceeded the practice of mimetic description. For Crane (and for Atherton and Dreiser), telling the truth was an imperative that demanded embodied, empirical knowledge. Both Crane and Atherton visited the electric chair in person before writing about it.[20] Thus, while each of them engaged with questions of representation, each also aspired to challenge the immateriality of reports that described the electric current as instantaneous and elided the social and legal apparatus that delivered the charge. For that reason, their creative narratives discuss material cultural more directly than most accounts by technical and popular-science writers at the time.

COMPLICATING THE NARRATIVE

Crane was not the only writer to complicate the fantasy of instant, easy-to-operate death. It was challenged by any writer who considered the broader contexts of electric execution. For example, after the Electric Execution Act was signed into law but before it was implemented, *Scientific American* identified a flaw its earlier articles had overlooked: the electric chair would have a real human body in it. Without questioning whether anyone should ever be placed into such a circuit, the 1889 article "Death by Electricity—the New Law of New York" constructs the person condemned to death as a variable that had not been accounted for in previous calculations: "The struggles of the prisoner, by disturbing the position of the electrodes, may bring about the most deplorable results."

Whereas *Scientific American* once framed electric execution as "simple," it now qualified that assertion: "As it is now, far too many executions by hanging fail in the end of quickly killing with little suffering. But where the vastly more complicated mechanism of an electric plant is depended on, the possibilities of a failure are largely multiplied." Having conflated death by electrical apparatus with "death by lightning" for ten years, *Scientific American* acknowledged the gross inadequacy of this analogy. Recasting power-generation equipment as "complicated," the 1889 article noted the previously elided difference between naturally occurring lightning and a human-built electric execution apparatus. It also acknowledged the agency of condemned people who might resist their roles as passive participants in their own executions—an agency that could not be entirely undercut by inventing new mechanisms for execution.[21]

Despite this late-dawning realization that the electric circuit would be more difficult to manage than early proposals imagined, the notion of the easy-to-operate apparatus had already taken hold of the American cultural imagination. Even the U.S. Supreme Court could not believe that a process long touted as simple and instantaneous could be cruel. When William Kemmler, the first man sentenced to die by electric execution, appealed his sentence on the grounds of the Constitution's Eighth Amendment, the official court ruling *In re: Kemmler* (136 U.S. 436), decided May 23, 1890, stated: "we think that the evidence is clearly in favor of the conclusion that it is within easy reach of electrical science at this day to so generate and apply to the person of the convict a current of electricity of such known and sufficient force as certainly to produce instantaneous, and therefore painless, death." The court's assurance that a "sufficient" apparatus "is within easy reach of electrical science at this day" echoes the optimism of *Scientific American*'s earlier articles. The court's use of the word *easy* specifically comports with technical accounts that highlighted technical ease to diminish the moral complexity of capital sentencing.

As it turned out, the Supreme Court and proponents of the chair had underestimated the technical complexity of electric execution. Kemmler's death was notoriously horrific.[22] Reports of his execution describe the spectators reeling in pain as they observe the botched execution. The *National Police Gazette* described the first moments after the electric current had been turned on and off as follows: "The witnesses, all supposing the man was dead, crowded around the death chair. Somebody removed the electrode from Kemmler's head. Then something occurred that froze the blood in the spectator's veins, and, as one witness said, made him long that he should be struck blind, for he could not close his eyes or turn his head, and yet what he saw fairly made him dizzy with horror." This August 23, 1890 report, titled simply "Electrocuted," challenges the death penalty by undermining its supposed instantaneousness. It also transforms Kemmler's body into a sensational narrative, erasing him as a victim of state power and rewriting his suffering as evidence against the authority of science. Kemmler's writhing, bleeding body attests not to a human's agony, but to the imperfect circuitry of the chair and the attending doctors' inability to diagnose his death.[23]

A more utilitarian report written to an audience of electrical experts generally corroborated the less sensational elements of this narrative. This August 16, 1890 article in *The Electrical World*, "The Execution of Kemmler by Electricity," quoted Charles R. Barnes, "who had charge of the dynamo at the execution" as saying that "the execution of Kemmler was a decided failure, but with the proper precautions it could have been successfully done." Barnes suggests that the resistance of the human body in the circuit was not appropriately taken into account, and he suggests moving the dynamo nearer to the execution chamber. To avoid undesirable vibrations in future executions, he warns against placing the dynamo on a hardwood floor.

As Sarat has argued (2001, 62), these writing practices emphasize the spectator's suffering above the victim's, suggesting that the execution was problematic not because it "reminded [viewers] of the ferocity of the state's sovereign power over life itself," but because it was not invisible or easily palatable to a genteel audience. Although descriptions of many bungled hangings often seemed more drawn out and painful than Kemmler's botched electrocution, his execution was considered a greater failure because it "suggested that the quest for a painless, and allegedly humane, technology of death was by no means complete" (ibid.)—a problem we still confront today in the rare occasion that an execution makes the headlines.[24]

Nonetheless, Kemmler's mishandled execution was not enough to change the law or to eradicate the fantasy than an instantaneous electric death was attainable. The movement to abolish the death penalty could not regain the momentum of the previous generation because the transition from asking whether to how was accompanied by new ways to disavow the brutality of the practice. Some believed that accounts of the botched executions were overstatements made by a notoriously sensational press.[25] Others, including the executioner Edwin F. Davis, hoped that death could be rationalized as scientific knowledge of the human body improved. Davis designed an "Electrocution Chair" that would collect biometric data while executing criminals to improve the process in future iterations (U.S. Patent 587, 649).[26] Even William Dean Howells, a staunch critic, remained struck by the promise of instant electric death. Although he didn't want the United States to adopt electric execution, he integrated this invention into his utopian *Altruria* series years later. In his otherwise idyllic novel *Through the Eye of a Needle*, the threat of electrocution is used to keep workers in their place: "when they attempted to break out, and their shipmates attempted to break in to free them, a light current of electricity was sent through the wires and the thing was done" (Howells 1907, 197).

In novels, newspapers, and dime novels, the myth of an easy, instant electric death endured, despite ample evidence that electric executions were considerably more troubling than advocates anticipated. "In electricity, as in many other things," *Harper's Electricity Book for Boys* informed its young readers, "simplicity is the keynote of success; and from this little device to employ alternating current for ridding a house of an insect nuisance sprang the grim apparatus known as the 'death chair,' used in the execution of first-degree criminals in the State of New York. Many people think the mechanism for electrocution is a complicated one, but it is quite as simple as the Edison roach-killer. One pole is placed at the head of the criminal and the other at the feet" (Adams 1907, 309). One dime novel's cover even depicted a monkey operating an electric chair, making fun of the presumed ease of the process. The persistence of this fantasy confirms the transition from whether to how in popular culture: proponents of capital punishment could believe in the notion of instant death while acknowledging the gruesome realities of electric execution in practice, because they held that the apparatus could be incrementally perfected, much like other electrical devices that were available for consumer use.

EXPERT TESTIMONY: REPRESENTING ELECTRIC DEATH AFTER KEMMLER

No account exemplifies the resilience (and, in hindsight, the insufficiency) of the enduring instant-death fantasy more clearly than Alfonso David Rockwell's 1920 autobiography *Rambling Recollections*. Rockwell was an electro-medical expert who testified in the Supreme Court case *In: re Kemmler* that electricity should not be considered cruel and unusual punishment. He oversaw the electric execution of scores of animals in order to advise the state of New York on improvements to the electric chair after Kemmler's execution. Electric execution was sufficiently significant to this man's career and life that he devoted an entire chapter of

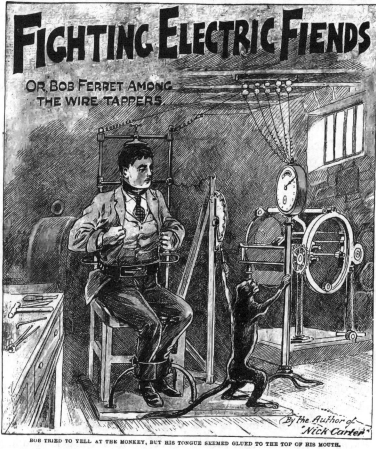

Figure 2.2
This cover of the *Nick Carter Weekly* depicts a monkey operating a home electric chair, adding a humorous and sensational twist to the fantasy that electric chairs were easy to operate and perfectible.

his autobiography to the subject. Nonetheless he perceived electric execution in the distant, romantic terms that Crane criticized.

Rockwell begins his chapter "Electro-execution" with a brief description of how sympathy shaped his involvement in the chair's implementation: "It had long been conceded that the rope was a barbarous method of execution, but it is always difficult to substitute a new method for an old, and the long contest over this *merciful* change in the law of the State of New York proved no exception. If the law must kill, let it kill decently. Although no strong advocate for capital punishment I *revolted at the brutality* of the strangulation method" (1920, 221, emphasis added). His "revolt" at this "brutality" reveals the hyperreality of hanging from his perspective: although Rockwell never witnessed "the horrors of hanging" firsthand, he trusted representations of the process completely.[27] Indeed, he describes it with some of the most vivid prose of his entire autobiography.

Rockwell recounts the process of hanging just as he must have encountered it in countless newspapers: "the first terrible fatal fall, the gradual choking, the blackening face, and protruding tongue, and above all the convulsive, agonizing and long-drawn-out struggle" (1920, 232). In contrast, his description of the four electrocutions he actually witnessed at Sing Sing Correctional Facility appears conspicuously less concrete: "[A]s one after the other, these miserable victims … took their places in the death-chair … a species of dreamlike apathy seemed to steal over me. It all seemed so unreal and without human touch that I could fancy myself wafted to the middle ages" (231). His disparate descriptions of electrocutions he witnessed and hangings he didn't witness reveal the indelible images of each execution method in the American cultural imagination. To Rockwell and to much of the American public, death by hanging was tangible. The horror and pain of asphyxiation was easy to imagine as it had been recounted in graphic, corroborative detail by journalists for decades. In contrast, death by electrocution appeared mysteriously intangible, occurring somehow outside of modern time (as Rockwell perceived it). Even this expert who helped modify the chair's hardware and witnessed electrocutions can only apprehend the experience as "unreal."

Rockwell didn't use this impression of unreality to absolve himself of responsibility. While he imagined the electric chair as an improvement from the brutality of hanging, he never perceived it as a perfectible solution. After describing the perceived "unreality" of the electric executions, he adds: "I experienced a feeling of shame and blood-guiltiness. As never before the awful meaning of the terms 'immutable' and 'irrevocable' was driven in. What if one of these men was not truly guilty?" (231). This chilling rhetorical question situates the electric chair in a larger social system. It suggests that even if the device could be improved dramatically, the process of condemning people to death would remain imperfect.

Rockwell concludes the chapter with a description of the electric chair that mimics the justification for its invention, though he qualifies this reasoning with a trace of sympathy:

> Aside from the knowledge that a human life is being sacrificed, there is nothing revolting in the sight. With face covered and person securely bound, the victim awaits the final stroke, and the translation from life to death is quicker than thought, and with the mathematical impossibility of pain. ... The certainty that no pain can be experienced under a lethal dose of electricity is evident from the fact that, while nerve force travels at the rate of but 100 feet a second, electricity travels at the rate of 160,000 miles a second. The brain, therefore, can have absolutely no time to experience a sensation, since the electrical current travels a million times faster than the nerve current. (Rockwell 1920, 232)

This passage mirrors *Scientific American*'s articles on the subject, drawing on Helmholtz to assert that electricity could kill before the nerves could transmit impulses of pain. Yet Rockwell modifies this assertion with the provocative phrase "Aside from the knowledge that a human life is being sacrificed."[28] This phrase quietly hints that the rationalized discourse of instant death cannot sufficiently justify the practice. If the scene can be perceived as clean and scientific only when the knowledge of death is set aside, then the idea of a purely scientific death is paradoxical. For this electro-medical practitioner, the question of how could not eclipse entirely the question of whether.

Ultimately the organization of the autobiographical chapter suggests that Rockwell understood electrical execution as a complex process that could be portrayed only by amalgamating different types of narration. He blended science and sympathy to describe the theory of electric execution and "the mathematical impossibility of pain." He used the language of fantasy and unreality to describe the moment of electric death. And he described the practice of capital punishment as haunting, despite the modernization of the practice. Like Twain's Yankee, Rockwell could account for this electric apparatus only by weaving together different perspectives and narrative modes, marrying his sense of modernity with an impression of the middle ages. He was not alone in that respect. Most articles about the chair stitch together scientific and sympathetic data in unsynthesized subsections.

Consider the newspaper coverage of the execution of Chester Gillette, on whose case Dreiser would base his novel. The two following passages are from the same newspaper article: "Chester Is Executed: Admits to Spiritual Advisers He Killed Former Sweetheart," published in the March 30, 1908 issue of the *Washington Times*. Their dramatic change in tone exemplifies the schism between quantifiable and emotional data that we also saw in Rockwell's autobiography. The first of these two passages reads:

> Chester slept about three hours last night. He seemed to be consoled most by the assurance of his spiritual adviser that trust in God was able to sustain him until the end. Chester entered the death chamber at 6:12:35 o'clock. He was strapped complete at 6:13:50. State Electrician Davis threw the switch at 6:14:03. The current was retained until 6:15:06. The execution was one of the quickest on record. A current, gradually raised to 1,800 volts, was retained one-half minute, and then reduced.
>
> With the quiver of the body, life passed out. The exact time that the execution was declared was 6:18 o'clock.

The second reads:

> At the sight of the ghastly scene, Chester rushed forward and took the chair. The cords of his neck swelled to abnormal size, his chest heaved with strong emotion, and on his forehead the beads of perspiration could be seen plainly. The legs of his trousers, slit to the knee, flapped ominously against his bare skin. The little round spot that had been shaved on the top of his head shone out like the roll of a tonsured priest.

The first passage dispassionately catalogues measurable details, such as the time of death and the voltage used. This style of reporting was common. In fact, other accounts, such as the March 30, 1908 article "Confession by Gillette Before He Died In Chair" in *The Evening World*, also mentioned the amperage—in this case, 7.5. Equally common was the nearly

Gothic style exemplified by the second passage reproduced above. The modifiers *ghastly* and *ominously* foster an eerie tone that stands in stark contrast to the earlier account that touted the event as "one of the quickest on record."

Atherton's subject, Carlyle Harris, was represented with similar heterogeneity. *The World* ran a full-page article and an illustration of Harris in the electric chair. The first column offers the technical details in large, well-spaced print: "The electric current that killed Harris was turned on at 12:40:30. The current was left on exactly 55¾ seconds." As in the examples I discussed above, the passive verb construction implies that the electrical current—not a person or disciplinary institution—was responsible for killing Harris. After unemotionally accounting his death, the article goes on to paint a sensational scene: "crowds gather" anxiously awaiting the news of Harris's death, while Harris's "fellow murderers" spend the entire night before his execution writhing in fear. The report takes up the entire front page, offering details that complicate the simple technical account of Harris's moment of electrocution.

These articles typify the rhetorical strategies used to report electric execution after Kemmler's execution and before World War I.[29] They privilege data about the moment the shock passed through the body, announcing this scientific information in larger print before layering in more sensational elements to their story. Perhaps these typographic distinctions attract different readers: the large print might attract the eye of the busy utilitarian, while the smaller print might spin a more sensational story for readers seeking entertainment. In either case, reports of electric execution often began in the near present with a quick report of the execution, then jumped back in time to describe the crime, the moments leading up to the execution, and finally the crowd's reaction before and after the "shock." This unsynthesized blending of styles lent a hybrid aesthetic to electric execution narratives, quite unlike the visceral descriptions of hanging that we saw in Rockwell's account. This fragmented style of reporting electric executions without regard for chronology or consistency added to the device's ambiguous characterization. It depicted legal electrocution—like the energy that enabled it—as simultaneously material and ethereal, scientific and sensational, fascinating and terrifying.

NOVELS ABOUT THE ELECTRIC CHAIR

Fiction was one cultural register that encouraged readers to ponder the incoherencies that were apparent but weren't remarked upon in news coverage. In fact, it was unsatisfying press coverage of the above cases that drew Atherton and Dreiser to compose their novels in the first place: both wanted to challenge existing conventions for representing criminal justice and capital punishment. This motivation was especially personal for Dreiser, who turned to fiction writing only after he realized that he could not tell the truths that concerned him in journalism. As the professor of communication Doug Underwood argues (2013, 116), "Throughout his journalism career, Dreiser walked a fine line between fascination and disgust with the conventions of a news gathering process that did not let him tell the full story behind the social and economic inequities, the municipal corruption, and the hypocrisies of urban life that he sanitized in his copy." By fictionalizing actual cases, both Atherton and Dreiser challenged readers who were interested in sensational crime dramas to consider other ways in which crime and punishment could be understood.[30]

Despite the 29-year gap between the release of *Patience Sparhawk* and *An American Tragedy*, the two books portrayed the electric chair with remarkable similarity. Both Atherton and Dreiser based their novels on "true crime" stories. Both interwove generic elements of the *Bildungsroman* (coming-of-age novel), criminal melodrama, and literary realism. Both also situated their concluding electric-chair sequences within broader contexts, tracing the development of their protagonists' lives, exploring how inequitable social systems incentivized violence, and highlighting the imperfections of the criminal justice system and the indulgences of American journalism.

Patience Sparhawk and *An American Tragedy* also diverged in ways that are worth noting before I attend to their most convergent scenes. The former novel was published shortly after the state of New York repealed the clause of its Electric Execution Act that gagged the press. First printed by The Bodley Head, a middlebrow house which was known for its controversial work by and about "New Women," *Patience Sparhawk* raised groundbreaking questions about the journalistic culture that profited from electric execution. Advertisements for the novel encouraged sensational expectations. A review in the October 1, 1897 issue of *The Dial* described the book as the "American maelstrom," and a "Topics of the Times" item in the May 15, 1898 *New York Times* quotes Atherton's claim that she "cannot write an article for a newspaper, much less a novel, without throwing the entire United States press into a ferment." Atherton thus invited readers to expect a sensational or controversial conclusion, then—as I will argue below—she cleverly confounded those expectations.

Almost thirty years later, Dreiser's novel was also widely anticipated. As his first publication since 1916, the novel was long awaited by the literary set (Pizer 2011). Dreiser's publisher—the highbrow, prestigious house Boni & Liveright—heightened those expectations by advertising *An American Tragedy* as the work of "America's foremost living novelist." The notion that this ambitious artist had been laboring over his masterpiece for nine years likely attracted readers with different expectations than those who enjoyed Atherton's novel. Nonetheless, Dreiser's work created even more of a stir than Atherton's. It was banned in Boston in 1929 for its allegedly perverse subject matter. The court case that ensued drew attention from newspapers across the United States when Clarence Darrow (a star witness) unsuccessfully defended the book's artistic merits. Accounts of the trial and of the novel both reveal that Dreiser's realistic account of crime and punishment in America was seen by some people as bold and risqué, though others (including the author's former friend H. L. Mencken) criticized it as bloated and boring.

These novels also differed somewhat in content and stated intention. Atherton wrote hers to poke fun at the culture of sensationalism, whereas Dreiser sought to reveal patterns of injustice in American society. To attain these goals, both writers employed the familiar rags-to-riches story line. In Atherton's novel, we first meet young Patience (the heroine) in her family's beautiful California Mission home—an idyllic setting marred by her mother's alcoholism. The first four books of the novel follow Patience as she develops more positive relationships with other female characters, who help Patience leave her abusive home and become a journalist for a newspaper called the *Day*. In the fifth and final book (the one that most concerns me here), Patience is wrongfully convicted for the murder of her husband, strapped into the electric chair, and astoundingly rescued moments before her electrocution.

Dreiser complicates his protagonist's rags-to-riches trajectory. Clyde Griffiths is a feckless, lower-class boy who moves from Kansas City to Chicago, then to the fictional city of Lycurgus, New York. Alienated from his wealthy extended family in Lycurgus, he pursues an affair with a woman of his own class, Roberta. When he finds the opportunity to court Sondra, a wealthy young woman with ties to the electrical appliance industry, Clyde imagines how much better his life could be without the nuisance of the now-pregnant Roberta: "If only—if only—it were not for Roberta now … . But for that, and the opposition of Sondra's parents which she was thinking she could overcome, did not heaven itself await him? … Once he and she were married, what could Sondra's relatives do? What, but take them into the glorious bosom of their resplendent home at Lycurgus or provide for them in some other way—he to no doubt eventually take some place in connection with the Finchley Electric Sweeper Company" (Dreiser 1925, 460).[31]

In the latter passage, and repeatedly throughout the novel, Dreiser depicts social climbers as employees of the electrical industry, and he describes wealthy homes as full of lavish electric light.[32] Sondra represents more than love to Clyde; she embodies the opportunity to be connected, socially and electrically. After Clyde meets Sondra, he plans to kill Roberta to free himself from his lower-class entanglement. But the moment Roberta dies appears so chaotic that even he doesn't know whether he committed a murder or witnessed an accident. Whereas Atherton characterizes Patience as innocent, Dreiser depicts his protagonist as ambiguously guilty.

The uncertainty of Clyde's guilt lends pathos to his execution. If readers might be indifferent toward the fate of the "guilty," as Schwed argued in the passage quoted above, Dreiser would frustrate their attempt to affix a definitive, ontological label of guilt or innocence onto a complex human life. In *An American Tragedy*, "guilt" is explicitly a legal status, divorced from the unwieldy, unknowable Truth. Dreiser depicts guilt as a human construct much like the electric chair. Both inventions rationalize the paradigm shift from whether to how.

Regardless of Clyde's "real" guilt or innocence, his experience adds a reproachful tone to the traditional story of class uplift. As Dreiser explained in a 1935 article about his real-world inspiration for the novel, "this was really not an *antisocial* dream as Americans should see it, but rather a *pro-social* dream. He [Clyde] *was really doing the kind of thing Americans should and would have said was the wise and moral thing for him to do had he not committed a murder*" (quoted in Pizer 2011, 32).[33] In other words, Clyde's desire to extricate himself from "Miss Poor" (as Dreiser dubbed Roberta's "type") exemplifies the values of his era—and Dreiser hoped that this fact would raise questions about those very values.[34] The final book of *An American Tragedy* effectively challenges Clyde's naive investment in the idea of social climbing by inverting the symbolism from earlier books in the novel. When Clyde finds himself living under Sing Sing's incandescent lights rather than Sondra's, he finds that electric lights don't necessarily signify wealth and warmth: "by incandescent lamps in the hall without at night—yet all so different from Bridgeburg—so much more bright or harsh illuminatively."

Unlike previous narratives about the electric chair (including those about Chester Gillette's electrocution, which served as an inspiration for it), *An American Tragedy* challenges the idealization or mystification of electric power. Dreiser even treats the electric shock anticlimactically, electing not to end the novel with his protagonist's death. After the electrocution, the narrative follows Clyde's priest as he witnesses the execution and wanders, dazed, into the street. It then turns to describe another poor family moving to the city. Echoing the language of the novel's introduction, these nested conclusions hint at the cyclical nature of Clyde's "tragedy": the poor repeatedly strive for the electric lights of the wealthy neighborhoods, but, lacking education and opportunity, they are more likely to find themselves in the electric chair.

JOURNALISTS IN THE ELECTRIC EXECUTION CHAMBER

Ultimately *Patience Sparhawk* and *An American Tragedy* share the conviction that electric death is not as instantaneous as most journalists and science writers might have readers believe. By taking time to land their protagonists in the electric chair and by emphasizing the material practices and human decisions that structure its use, each narrative contends that the chair's meanings are laboriously produced around the moment of electric shock. To convey this point, both emphasize the role of journalists in shaping the execution process.[35]

In *Patience Sparhawk*, this theme becomes apparent when a reporter from the *Eye* (rival paper to the heroine's employer, the *Day*) becomes the first to accuse Patience of murder. After inspiring her wrongful conviction, journalists eerily influence Patience's behavior in the execution chamber. In fact, throughout the novel's electric chair sequence, Patience's awareness of herself as news overshadows her fear of imminent death. Drawing upon the language of "unreality," the narrator describes the protagonist's inability to grasp that she has been sentenced to death: "there was a sense of unreality in it all. She felt as if about to play some great final act; she could not realise that the climax meant her own annihilation" (Atherton 1897, 481).[36] This passage complements Crane's "Devil's Acre" in that it illuminates the stakes of perceiving electric execution in trite, romantic terms. As long as Patience interprets the electric execution chamber as a setting in the journalists' sensational narratives, she abandons her sense of agency. Although she had been a willful, outspoken character in the earlier books of the novel, Patience suddenly contributes to "her own annihilation" by choosing to act like a storybook hero. In so doing, she expedites her own execution process while the prison officials try to delay it; she hopes to be portrayed honorably while they hope for a stay of execution.

The protagonist of *An American Tragedy* feels no such pressure to act heroic, but his experience similarly exhibits how sensational journalists corrupt the criminal justice process. Rampant public interest in Clyde's death inhibits the possibility of a fair trial. The jury cannot be impartial. Only one juror, Samuel Upham, admits to sympathizing with the defendant, and his opinion is easily swayed by a threat of publicity: "[H]e was threatened with exposure and the public rage and obloquy which was such to follow in case the jury was hung. 'We'll fix you. You won't get by without the public knowing exactly where you stand.' Whereupon, having a satisfactory drug business in North Mansfield, he at once decided it was best to pocket this opposition to Mason [the district attorney prosecuting Clyde] and agree" (792). This quotation makes the corruption of Clyde's trial conspicuous, although its lack of attribution diffuses the blame. Dreiser's characteristic opacity inhibits readers from accusing any single corrupt individual for this violation of his protagonist's rights. Instead the scene encourages readers, as they wonder who threatened Upham and why, to consider how many interested parties might unduly influence a court's decision.

Dreiser emphasizes the corrupting influence of sensationalism by insisting that it saturates Clyde's entire courtroom experience with "the strong public contempt and rage that the majority of those present had for him from the start" (774). Publicity about the case poisons the jury pool and sets public opinion against the defendant. The novel distills the public's seething hatred for Clyde in one outcry from a courtroom spectator: "Why don't they kill the God-damned bastard and be done with him?" (776). This interjection from an unaffected party raises a question that both novels explore: How does the public become invested in the execution of a stranger? Certainly the state desired this response; the move toward "scientific" and "humane" execution methods was intended to preempt the crowd from sympathizing with the condemned party.[37] Nonetheless, in these novels the crowd's titillation undermines the supposed advancement and humaneness of the execution. *Patience Sparhawk* and *An American Tragedy* both imply that the process of assigning guilt and influencing public opinion must be considered a part of the apparatus of electric execution—and that these components could never be considered "easy" or "instantaneous."

Both novels characterize the sensational press as part of this problem, and in the process both implicate readers as well as writers. In *Patience Sparhawk*, Atherton raises questions about journalists' and readers' voyeuristic interest in following a crime story to its most sensational conclusion—in this case, the electrocution of an innocent woman. In a scene that could have been taken verbatim from news coverage of Carlyle Harris's death, the narrator describes a crowd of women gathering to await the news of the protagonist's execution: "The women sat about on the slope opposite the prison, pushing the baby carriages absently back and forth, or gossiping with animation. Other women crowded up the bluff, settling themselves comfortably to await, with what patience they could muster, the elevation of the black flag" (Atherton 1897, 479). Since the novel's earlier books underscored the importance of fostering strong relationships among women, these curious mothers represent a domestic femininity that seems somewhat corrupted by their desire for Patience's electrocution. Although the women behave in accordance with the law, their collective interest comes across as misdirected—especially because the reader knows Patience to be innocent.

Dreiser includes a resonant image of perverted sentimentality in *An American Tragedy*. In a moment of particular pathos, the narrator recounts: "On the following morning Clyde was arraigned for sentence, with Mrs.

Griffiths given a seat near him and seeking, paper and pencil in hand, to make notes of, for her, an unutterable scene, while a large crowd surveyed her. His own mother! And acting as a reporter! Something absurd, grotesque, insensitive, even ludicrous about such a family and such a scene" (809). Dreiser, like Atherton, emphasizes how the relationship between journalists and criminal justice can degrade sympathy. In other words, they suggest that the repackaging of executions as entertainment displaced ethical questions and replaced them with technical and socioeconomic ones.

Whereas Clyde's mother signifies the nefarious influence of sensationalism, the voyeuristic women in *Patience Sparhawk* function as avatars for the reader's experience. As the scene shifts from the women ravenously awaiting the news of Patience's death to the execution chamber, Atherton encourages readers to anticipate her protagonist's electrocution with similar excitement, synthesizing their perspective with that of the women on the hill. By painting Patience as a non-religious, outspoken heroine in the earlier books of the novel, she urges readers to wonder whether this unconventional character should be rescued. Since Patience had fallen in love with her lawyer shortly before being condemned to death, the possibility of her execution enables the classically tragic element of nearly missed love, and it offers a description of something that had not yet happened anywhere in the world: the electrocution of a woman.[38] Indeed, a review of the novel in the April 15, 1897 issue of *The Chapbook*, titled "The Epic of the Advanced Woman," said that Patience's rescue comes "somewhat to the regret of the sympathetic reader."

An April 24, 1897 review in *The Critic* expressed disinterest in the love story between Patience and her lawyer, Garan Bourke, while admitting that the novel "closes with an excellent report of a murder trial, in which there is an unmistakable touch of dramatic power." Such statements were common in criticism of *Patience Sparhawk*, demonstrating that many readers were more invested in the criminal than the romantic plot although both subgenres were highly popular at the time. This critic's heightened interest in the criminal plot suggest that such readers' experiences may coincide with those of the women on the hill, as they eagerly await news of Patience's electrocution.

Whether readers hope for Patience's demise or root for her release, Atherton's rhetoric urges the reader to anticipate the protagonist's electrocution. She draws out the final scene, amplifying the suspense:

> [Patience] dropped the head-keeper's arm and walked deliberately to the chair; but he caught her hand and held her back.
> "Wait a minute," he said, with affected gruffness. He went to the chair and examined it in detail. He asked a number of questions, which were answered by the electrician with haughty surprise. In a moment the reporters were staring, and like a lighting flash one brain informed another that "something was in the wind." (Atherton 1897, 484)

The head-keeper's questions delay the novel's progression and intensify its climax. They also call to mind the frequency of mistakes associated with the electric chair and invoke familiar electrocution narratives by reminding the reader that any of the chair's mechanisms could malfunction. Throughout this scene, Patience's matter-of-fact approach to her death and the narrator's use of images that evoke electricity ("like a lighting flash") emphasize the seeming inevitability of her electrocution. At this point in the novel a botched execution seems more likely than a rescue—especially since the climax occurs on the last page. The strategic placement of this scene encourages readers to imagine that the book will end with the death of its eponymous protagonist.

In *Patience Sparhawk,* Atherton heightens the reader's curiosity for the "shocking" conclusion by repeatedly describing sensations that *seem* to portray the moment of electric shock: "Her mind was a sudden blaze of light—white light she thought with a stifled shrink—in which every detail of the room was sharply accentuated" (1897, 486). This "blaze of light" reads like an imaginative account of the sensation of electrocution, but it merely describes Patience's fear. After building readers' excitement for Patience's death in this way, Atherton discharges the suspenseful energy in a suddenly romantic turn. On the last page of the novel, Bourke rescues Patience just as she braces for the electric shock:

> Suddenly her ears were pierced by a din which made her muscles leap against the straps. Was she in hell, and was this her greeting? She had felt a second's thankfulness that death had been painless.
> Then, out of the babel of sound, she distinguished words which made her sit erect and open her eyes, her pulses bound, her blood leap. ... The cap had been removed, the men were unbuckling the straps. ...] Round her the newspaper men were pressing, shouting and cheering, trying to get at her hand to shake it.
> She smiled and held out her hand, but dared not speak to them. Pride still lived, and she was afraid she should cry. (488)

This final scene opens as Patience's muscles "leap against the straps," tantalizing readers with the promise of electrocution. By coercing her readers to imagine that the electrocution is over before revealing that the current hasn't even been switched on, Atherton pokes fun at the fantasy of instant and painless death. And by short-circuiting the electrocution narrative, she forces readers to confront their fantasies about the fatal spark and their perverse desire to see the sensational narrative through to its end. On this final page, the protagonist's name takes on new meaning: Atherton rewards Patience (both the virtue and the character), while she calls into question the desire to rush to a "shocking" conclusion.

Atherton also refuses to let the romantic turn entirely subsume her conclusion. Even during her rescue, Patience continues to restrain her own behavior in the presence of journalists. She refuses to swoon or to act the part of the damsel. In this way, Atherton illuminates just how mediated the execution chamber must be. Ultimately her mélange of romance and sensationalism suggests that readers' interest in the mystery of electric death implies their complicity in the practice. Journalists, audience members, and even the heroine herself would have let Patience die for the sake of a great story. But Bourke, with his blend of sentimentalism and legal pragmatism, refuses to let this story play out according to its expected melodramatic conventions.

It may seem unusual for Atherton to play with her readers in such a way, but the novel's form is consistent with her self-identification as an author of unconventional intelligence and honesty. Atherton aspired to represent the "real," although she disliked the practitioners of literary realism and considered the genre myopic and faddish, famously referring to it as "littleism."[39] She wanted her books to be popular, but she also wanted them to be exceptional in their representations of women and romance—aspects of life that she felt William Dean Howells's form of literary realism failed to address adequately. The form of *Patience Sparhawk* represents Atherton's first attempt to deal with these issues by combining popular modern themes (crime, electric execution) and a romantically inflected form of realism that attempts to capture human emotion mimetically.

Twenty-nine years later, Dreiser improved upon the psychological realism to which Atherton aspired. By the time *An American Tragedy* was published, the discourse about electric execution had changed appreciably. After a world war, the execution of individuals seemed less sensational. The rise of systems thinking also colored Dreiser's interpretation of capital punishment. Between the publication of *Patience Sparhawk* and *An American Tragedy*, Thorstein Veblen created a new lexicon for his cultural moment: he identified the rise of "conspicuous consumption," and he began to use the word *technology* to describe what engineers actually produced—new knowledge. Dreiser was not an early adopter of the word *technology* as Veblen used it, but both writers shared an interest in re-imagining America in systemic terms.[40] Dreiser spent more than ten years cultivating a strategy for describing the interconnectedness that he intuited in modern American life. This ambitious desire to describe realistically the system and the individual's place within it was his lasting legacy in American letters, influencing other writers who came after him to do the same.[41]

Since his novel was written in this distinctive historical context, Dreiser didn't invoke, as previous writers had, the discourse of electricity as "unreal." Instead, he demonstrated that the cruel and unusual aspect of electric execution doesn't inhere in the chair itself; it emerges from the psychological duress of waiting in expectation for an unimaginable death.[42] As Crane argued in "The Devil's Acre," the chair is horrible because it "is patient—patient as time." After the court sentences him to death, Clyde ponders that patient chair with horror: "And then to sit in that chair he had seen in his mind's eye for so long—these many days and nights when he could not force his mind to drive it away. Here it was again before him—that dreadful, ghastly chair—only closer and larger than ever before—there in the very center of space between himself and Justice Oberwaltzer. He could see it plainly now—squarish, heavy-armed, heavy-backed, some straps at the top and sides. God!" (794). The contrast between Clyde's psychological horror and his mundane mimetic depiction of the "squarish" chair with "some straps" accentuate his mounting anxiety. The dashes accentuate the character's mental anguish. He can think of nothing but the uncannily commonplace object that will kill him.

In prison, Clyde's dread of the chair intensifies. He lives in a state of constant awareness that his cell is connected by corridors and copper wires to the device that will eventually kill him. The narrator describes Clyde's mental state with a staccato of short sentences that elicit an impression of his racing pulse: "His nerves were as taut as cords about to snap. … That other room! It was in here somewhere too. This room was connected with it. He knew that. There was a door. It led to that chair. *That chair*" (815). Throughout this prison scene, the word *connected* comes to seem more ominous, as the proximity of the electric chair tortures Clyde before he is ever strapped into it: "And then—as in that dream in which he turned from the tangle of snakes to face the tramping rhinoceros with its two horns—he was confronted by that awful thing in the adjoining room—that chair! That chair! Its straps and its flashes which so regularly dimmed the lights in this room" (841–842). The imagery of the frequently dimming lights, based like much of the novel in verifiable fact, reads as sinister because it deviates starkly from the novel's previous depictions of electric light as a symbol of wealth and class uplift. It also invites the reader to wonder how often incarcerated people are electrocuted. Perhaps Clyde's perception of these dimming lights was exaggerated by his fear of his own death sentence; perhaps electric execution has grown hauntingly commonplace.

Dreiser's distant, omniscient narrative voice underscores the psychological torture of Clyde's experience and the absurdity of the entire disciplinary process. From this point of view, the narrator destabilizes the discourse of "scientific progress" that *Scientific American*, Alfonso David Rockwell, and other proponents championed. He characterizes the prison system as irrational and inhumane:

> The "death house" in this particular prison was one of those crass erections and maintenances of human insensitiveness and stupidity principally for which no one primarily was really responsible. Indeed, its total plan and procedure were the results of a series of primary legislative enactments, followed by decisions and compulsions as devised by the temperaments and seeming necessities of various wardens, until at last—by degrees and without anything worthy of the name thinking on any one's part—there had been gathered and was now being enforced all that could possibly be imagined in the way of unnecessary and really unauthorized cruelty or stupid and destructive torture. (815)

The narrator describes incremental changes to prison spaces—often associated with the ideology of progress—as exacerbating the condemned person's traumatic experience. Since no person takes responsibility for the institution or the people confined there, Clyde inhabits a space that has been entirely shaped by disparate whims, "human insensitiveness," and "stupidity," all of which intensify his experience of tortured waiting.

Although "no one was really responsible" for the prison's severity, Dreiser suggests that someone should be accountable. The narrator depicts Clyde's waiting as a violation of the Eighth Amendment, and he attributes the cruelty and unusualness of the process to the rise of mechanistic systems that lack human oversight and emotion. Accounting for the prison experience in this way, Dreiser implies that a human touch in this space could improve the conditions for the people who work and live and die there: "There was a system—a horrible routine system—as long since he had come to feel it to be so. It was iron. It moved automatically like a machine without the aid or the hearts of men. These guards! ... they were iron, too—mere machines, automatons, pushing and pushing and yet restraining and restraining one" (866). This passage characterizes the diffusion of responsibility as inherently dehumanizing. Even the prison guards lose their humanness within these walls.

Dreiser uses mechanistic imagery in these scenes, but he paints electrocution as a human rather than an electrical invention. His execution scene omits the bright lights and muscles leaping of Atherton's account, and it forgoes the calculated details of journalistic reports. Dreiser instead builds anticipation by emphasizing the human processional: "the guards coming, first to slit his right trouser leg for the metal plate and then going to draw the curtains before the cells. ... And then the final walk with the Reverend McMillan on his right hand and the Reverend Gibson on his left—the guards front and rear" (869–870). The dashes become more frequent as Clyde experiences the dreaded sense of unreality that other observers had described. However, this character attributes the impression of disbelief to his own body and voice rather than to the mysteriousness of the "death spark":

> And Clyde, with enough earthly thought and strength to reply: 'Good-by all.' But his voice sounding so strange and weak, even to himself, so far distant as though it emanated from another being walking alongside of him, and not from himself. And his feet were walking, but automatically, it seemed. And he was conscious of that familiar shuffle—shuffle—as they pushed him on and on toward that door. Now it was here; now it was being opened. There it was—at last—the chair he had so often seen in his dreams—that he so dreaded—to which he was now compelled to go. He was being pushed toward that—into that—on—on—through the door which was now open—to receive him—but which was as quickly closed again on all the earthly life he had ever known. (870)

In this scene, the narrator downplays the allure of the chair and the electric current. He implores readers to recognize that human life—specifically a person's ability to approach his own death with a modicum of bravery—is more amazing than any apparatus could be. The narrator emphasizes the singularity of Clyde's experience with the repetition of the monosyllablic word *now*. This word confers a certain power and knowledge to Clyde that remains inaccessible to everyone else—to reader and writer, to electrical expert and prison guard. It draws attention to the moments that seem conspicuously absent from past-tense reports of trials and executions: only Clyde experiences this proceeding as a *now*. His last *now*. This unknowable experience is Clyde's alone. The narrator never describes the protagonist's impression of the electric current.

This series of *nows* distills the rhetoric of instantaneousness that was so prominent in other accounts of electric execution. After a lengthy passage about instants, Dreiser summarizes the actual electrocution with a single adverb—*quickly*—though the expectation and performance of execution had been anything but quick. Readers may perceive this depiction as corroborating Helmholtz's claim that the electricity would kill before the nerves could transmit the sensation of the current coursing through them. Nonetheless, the narrator consistently places the emphasis on Clyde rather than the execution method. This strategy suggests that the experiences of fear, poverty, and the inevitably failed pursuit of wealth are each more noteworthy aspects of the execution apparatus than the electrical current itself.

THE STAKES OF REPRESENTATION

Taken together, Dreiser, Atherton, and the anonymous journalists I discussed above can all be said to emphasize unreality and fragmentation, qualities that T. J. Jackson Lears associates with an anti-modernist perspective (1983, 3–32). But these texts are too multifarious to advocate only one viewpoint. They amplify concerns that lie latent in accounts such as Rockwell's autobiography by asking whether any execution method could ever be wholly scientific or perfectible. These texts complicate the common practice of describing the electric chair in terms of its adjective. They insist that this device is more than just "electrical" or the connotations that attend this word (instantaneous, modern, natural); it is always and inextricably defined by its relationship to other systems (the power grid, the sensational press, the class system).

Dreiser and Atherton encourage readers to question why they invest in the execution of people they do not know. They also invite readers to consider how their own lives are entangled with the very systems that lend (legal, cultural, and electrical) power to the "death chair." Both insinuate that the practice of capital punishment cannot be reduced to a narrative about scientific or humane technique of execution. A more accurate story about the death penalty must also consider how our newspapers turn individuals who have made mistakes into monsters, reducing their life stories to a sequence of warning signs that could not have been obvious amid the noise of living. Still, we should not overstate the interventions of these relative outliers. Even if Dreiser's and Atherton's accounts of electric execution can help us identify previously overlooked patterns in the history of capital punishment, they didn't rekindle the death-penalty abolition movement or perceptibly change the dominant discourse about this practice of state power. What, then, did these novels accomplish? How did they remain in print without inciting action?

As they strove to re-contextualize the electric chair, Dreiser and Atherton both staged a battle between individuals and systems. They cultivated a mood akin to Leo Marx's "Postmodern Pessimism," although they pre-dated the rise of this aesthetic and theoretical movement: both developed characters who fret about their "diminished sense of human agency" in a world of large-scale systems that seem to have their own unstoppable momentum (Marx 1994, 257). The success of these novels emerged from their gripping and relatable depictions of overcivilization and discontent. Each sounded an elegy for the individual's diminishing autonomy. This concern for the place of the individual in a networked world was what allured readers for generations; this concern eclipsed the other political issues that each novel raised.

If Dreiser and Atherton didn't incite the social change they both desired, they did at least lend their voices to the chorus of writers who began to invoke electricity as a metonym. The electric chair functions in their narratives as both a historical referent and as a polysemous symbol for the desirable and undesirable connections that affect the course of a human life. At an allegorical level, that means that Bourke's rescue of

Patience celebrates the possibility for individuals to change the momentum of a system—a hope that Dreiser, writing decades later, could no longer share.[43] Of course, as we will see, not all writers who employed electricity as a metaphor saw "systems" as external and threatening to individuals. Even *Patience Sparhawk* hints at the potential for a more affirmative reading of electrical development.

Before the crime-and-punishment storyline of the fifth book of *Patience Sparhawk*, the heroine's rise from an abused troublemaker into an independent woman coincides with her growing sensitivity to electricity and magnetism. In fact, Atherton (1897) weaves vivid images of electricity throughout the entire novel. Every key moment in Patience's life is marked by her heightened awareness of electrical and magnetic forces: "She felt higher above the earth than ever before, but more conscious of its magnetism" (36). "Her feet had touched that nether world where the electrical forces of the universe appear to be generated, and its wonder—not the man—conquered her. She shook horribly" (177). "There seems to be some tremendous magnetic force in the Universe that makes the human race nine-tenths Love—for want of a better name" (184). This underlying metaphoric structure hints that electricity has much more to offer the American public than a faster form of execution.[44] By describing the untapped potentials of electric and magnetic fields, Atherton hints that Americans should focus on harnessing this energy in life-affirming ways even before she satirizes its life-threatening applications.

This longing to foster the positive connotations and uses of electricity proliferates across American letters—especially in the utopia, a genre I discuss at length in chapter 3. Nonetheless, as we saw in *A Connecticut Yankee*, the progressive and destructive aspects of electricity were not so easily extricable. These fears and fantasies evolved in feedback with one another. In this chapter I have chronicled how the electric chair came to symbolize the individual's loss of agency. But these fears about individual vulnerability didn't displace the fantasy that electric power might enhance or signify the individual's capacity for increased control. After all, the person flipping the switch experiences the electrical system differently than the frightened bystander who fearfully watches the lights dim. Today, we regularly adopt both perspectives. We overstate our own control over electrical systems. When we say "I turned on the lights" or "I sent you a message," we elide the work and the decisions of electrical experts and construction workers who wired our communities in the first place.[45] We simultaneously feel small in the face of large-scale systems, and we feel empowered by the interfaces we can control.

Before these conflicting (yet coincident) feelings became a tacit part of daily life, writers meditated on these plural potentials in ways we have forgotten—in ways we might productively revisit. If the electric chair narratives discussed above lend themselves too easily to our own forms of "Postmodern Pessimism," consider a final example that highlights these ambiguities, instead of trying to collapse them: a May 1, 1893 article in the Progressive periodical *The Independent*. The unassuming piece, titled "Touching The Button," invites the reader to "stand by the electric chair" and continues as follows: "There sits a man in his usual health. No cause appears; but suddenly he dies. A flash, as it were from the clouds, invisible, with no cause at hand, the sheriff somewhere else and unseen; and he dies." The next paragraph describes an uncannily similar scene: "The touch of the button by the President starts into active motion the ponderous machinery of the Exposition. Where was he? Invisible, somewhere else. When he touches the button, every wheel starts, every process of beautiful production goes on before our eyes." From these two buttons, the writer extrapolates a neo-Transcendentalist philosophy of interconnectedness. He calls these buttons "the lesson of human influence," asking "Which of us is not pressing the button? What we do here is seen and felt, invisible, far off." The article ends with a suggestive metaphor: "There is a button under every finger." The thought can be terrifying, as we have seen, but also potentially exhilarating—depending upon where the lines from that button might lead.

3 CHARLOTTE PERKINS GILMAN'S HUMAN STORAGE BATTERY AND OTHER FANTASIES OF INTERCONNECTION

The rhetorical techniques I discussed in chapter 2 are likely to have sounded familiar. The story of the alienated individual struggling to survive in a world of inimical systems remains popular today. Although we know this story well, we also know that it is not the only one we commonly tell about technology or electricity or the future. For generations, such cynical narratives about technology have developed in tension with an equal and opposite tendency toward utopianism. Whereas Dreiser feared the diminished agency of the individual in a network society, other writers hoped that emerging systems would offer increased opportunities for control over ungainly aspects of their world.[1] The Progressive Era activist Charlotte Perkins Gilman was one such utopian. Much like prognosticators of our own information age and other utopian or socialist writers of her era, she hoped that electrical systems and other emergent networks would promote profound social change.

Gilman's critics and biographers didn't indulge in the technological fallacy as often as Twain's critics. But that was a result of biases that were built into the word *technology* rather than of an especial attention to etymology. As Ruth Oldenziel argues (1999, 40–42), the concept of technology specifically evolved to marginalize conventionally feminine forms of innovation. Although present-day scholars have rarely considered technology as a dominant theme in her body of work, Gilman seems to have been engrossed with systems and their management. To Gilman's mind, modern sewer, transportation, communication, and power systems remapped the boundaries of the "domestic sphere." In her 1901 treatise *Concerning the Children*, she uses that idea to reframe the traditional home as a node in a complex of larger networks. In so doing, she experiments with a strategy that she would perfect in her most studied novel, *Herland* (published in 1915): she undermines prevailing gender conventions by making them seem ridiculous from an estranged but reasonable perspective. In *Concerning the Children*, as in *Herland* and her other aspirational pieces, Gilman reclaims the stereotype that women are guided by natural maternal instincts in order to promote alternative ways that these instincts could be harnessed:

> Like an ostrich with his head in the sand, the mother shuts herself up in the home and imagines that she is safe and hidden, acting as if "the home" was isolated in space. That the home is not isolated we are made painfully conscious through its material connections,—gas-pipes, water-pipes, sewer-pipes, and electric wires,—all serving us well or ill according to their general management. Milk, food, clothing, and all supplies brought in bring health or disease according to their general management. The mere physical comfort of the home needs collective action, to say nothing of the psychic connection in which we all live, and where none is safe and clean till all are safe and clean. (1901, 289–290)

In this passage, Gilman situates the modern home within a maze of pipes and wires to debunk the myth of its insularity. By alleging that these conduits can bring "health or disease," she implies that the truly attentive mother should be more concerned with the "general management" of municipal systems than with the administration of an arbitrarily atomized home. She also maintains that no individual could be responsible for this tangle of large-scale systems. "Collective action" seems the only appropriate response. Gilman hints here and argues explicitly elsewhere that modern Americans should resist thinking as individualists. She encourages her readers to re-imagine themselves in terms of the systems they use, regulate, and build. As I will explain below, electricity plays an important role in Gilman's discussions of this desired conceptual shift. Although scholars have focused primarily on her uses of evolutionary dicta, Gilman also worked thoughtfully with the discourses of electrical development. Drawing on electricity's double association with vitality and with industrialized modernity, she invoked it to claim that individuals in a system are not cogs in a machine, as per the idiom of the first industrial revolution, but rather are cells.[2] And she means that in more ways than one: they are electrical batteries and the basic units of life.

Before considering the electrical and vitalistic imagery in Gilman's body of work, let us pause to analyze the above lines from *Concerning the*

Children. While this passage imagines an enticingly expanded social role for women, it also appeals to conservative fears about public health and illness. The alarmist plea that "none is safe" registers the ineluctable links between the author's socialism, her feminism, and her racism: Gilman strives to manage systems so as to liberate certain women and to limit others. Recent scholars have cited similar passages to revise the feminist's sanitized legacy in our literary histories, urging critics to acknowledge both the salubrious and the sinister aspects of her social vision. A previous generation of scholars had celebrated Gilman's activism for her gender, reading her texts against the more famous and masculine utopias of Edward Bellamy and his disciples. In contrast, Jennifer Fleissner (2004, 91) has shown that Gilman's writing actually accords with the "technocratic, overtly militaristic" literature of her era. Katherine Fusco (2009, 424) agrees, claiming that we must study the writer's complex imaginary of systems if we strive to comprehend "Gilman's feminism and her racism, without one necessitating the exclusion of the other." I extend such conversations about the writer's legacy by arguing that Gilman subtly inscribes social power lines onto modern electrical ones—a feat she accomplishes by alternating between literal and figurative depictions of these systems.

Much as Mark Twain does in *A Connecticut Yankee*, Gilman fosters a hybrid understanding of electricity. In Twain's novel, this energy could signify naturally occurring lightning bolts or button-operated circuits, dazzling displays of light or blue sparks of death. In Gilman's body of work, electricity refers to all those things—but also to nerves and bodies. This additional set of referents was particularly unwieldy in Gilman's time because of the close correlation medical scientists drew between bodies and electricity. Tim Armstrong explains this phenomenon concisely (1998, 7): "If one consider the idea of the 'body electric' … it is almost impossible to clearly assign a metaphorical status to the term 'electric.'"

Electricity functioned as more than a metaphor because nineteenth-century physiologists used electrical equipment to understand how animal nerves and muscles worked. As the scholar of technology studies Steve Woolgar argues (2012, 304), this redefinition of the human with new electrical implements was indicative of a broader trend: "[D]ifferences in interpretations of technology both express and give rise to competing preconceptions about the essential quality of humans. In discussing and debating new technology, protagonists are reconstructing and redefining the concepts of man and machine and the similarity and difference between them."

This impulse to understand the body in electrical terms changed dominant conceptions of both terms. Electrical systems came to seem more sinewy: Gilman and many writers of her era construe power lines as "nerves" and "tissues." Bodies, in turn, seemed less wholly organic: from this intellectual tradition the field of bioelectricity was born, as were Donna Haraway's (1990) cyborg and N. Katherine Hayle's (1999) posthuman constructions of the self. Of course, *body* and *electric* are not the only "words whose intentions are fugitive," to borrow William Burroughs's phrase. The intermingled intentions of these particular words matter because they encouraged activists such as Gilman and characters such as Twain's Yankee to misrecognize "electrical progress" as human progress in general and electrical lighting as enlightening in the broader sense of the word. These entangled connotations allowed Gilman and other Progressive Era writers to graft their Darwinian biases onto their discussions of electrical development. In other words, it enabled them to claim that people who used electricity were inherently, physiologically superior to those who did not. In her body of work, Gilman uses this blend of evolutionary and electrical prejudices to carry the imperialist themes in *A Connecticut Yankee* to their logical extreme. Her treatises and her fiction code electrified societies as superior to—indeed entirely distinct from—societies that use fewer networks. This narrative strategy was not unique to Gilman; many writers of her era shared her presumption that electrical systems represented a step in human evolution.

Gilman's eclectic body of work can help us understand how electricity came to serve this politically charged rhetorical function in American letters. In fact, her catalog offers an especially interesting case study because she wrote across so many disciplines. As a writer of utopias, Gilman affords us an opportunity to examine the boom in that style of writing that occurred during her lifetime.[3] She also worked as a lecturer, a poet, a fiction writer, a sociologist, and an advocate for women's social and medical rights. In the process, she drew attention to the permeable boundaries between these fields by mixing metaphors and methods from different intellectual traditions.

Gilman also warrants such investigation because her aspirations typify those of other progressives and socialists of her era. Technicians and experts—most notable among them Charles P. Steinmetz, General Electric's famous electrical engineer—shared Gilman's conviction that electrical systems would augur social change. In fact, Steinmetz served on the editorial board of the *New Review* with Gilman. Though this fact says nothing about whether or not they actually met and exchanged ideas, it does indicate to some extent their convergent political interests. Like Gilman, Steinmetz believed that electrification was pushing American society beyond individualism and toward a presumably evolved state of interdependence. Even in technical speeches, he often found a way to promote his understanding of electrical systems as inherently socialistic. Consider the following excerpt from a speech about the state of electrical development that he delivered to experts during an industry retreat:

> The relation between the steam engine as a source of power and the electric motor thus is about the same as the relation between the individualist and the socialist, using the terms in their broadest sense; the one is independent of everything else, is self-contained, the other, the electric motor, is dependent on every other user in the system. That means, to get the best economy from the electric power, co-ordination of all the industries is necessary, and the electric power is probably today the most powerful force tending towards co-ordination, that is co-operation. (Steinmetz 1913, 70)[4]

If this passage calls to mind *fin de siècle* utopias about coordination, such as Edward Bellamy's *Looking Backward* (published in 1888) or Gilman's *Moving the Mountain* (1911), that may be because the setting of this speech had an aura of the utopian. It was presented during the proceedings of "Camp Co-operation," a gathering of electrical industry insiders held (on an island they owned) for the purpose of discussing the importance of "co-operation"—a term that meant monopoly building to some and social progress to others. The quoted passage stands apart from the rest of Steinmetz's speech as poignant and surprising. Equating steam power and individualism, electricity and socialism, Steinmetz sketches a timeline that casts electrified societies as more evolved and socialistic than others. Gilman would repeatedly and vociferously agree.

Steinmetz and Gilman were two voices amid a chorus of visionaries who imagined that electricity might foster social change. Works that conflated "electrical progress" and social progress bridged the gap between technical and literary writing, even as this divide widened in the face of increasing specialization. Indeed, while Gilman was writing across multiple genres and cobbling together elements from different intellectual traditions, practitioners in other fields were engaging in a similar range of writing practices. Alongside the utopian tract and the utopian novel, the "Edisonade" dime novel appeared. I will elaborate on that genre in chapter 4. Suffice it to say that these works celebrated how electricity, radium, x rays, and other energies could serve the public good—when controlled by the right people. Dystopian literature reciprocally explored the probability that these energies would fall into the wrong hands. Meanwhile, professors and practitioners of electrical science wrote creatively about their control of electric power. John Trowbridge, a professor of physics at Harvard University, wrote a series of books for young adults that combined conventional rags-to-riches story lines with detailed descriptions of electrical circuits.[5] In 1919, Charles M. Ripley, a factory worker at General Electric, published *A Romance of a Great Factory* (with an introduction by Steinmetz), which described the electrical factory of the time as a quasi-utopia. A wide array of *fin de siècle* texts—including the "literary" writing of Bellamy and Gilman, the lowbrow Edisonades (many of them ghost-written for publishing syndicates), and the genre-defying writing of electrical experts who celebrated their profession in prose—conveyed excitement about the new possibilities for reformist social control that electricity seemed to offer. Although in this chapter I focus primarily on Charlotte Perkins Gilman's multidisciplinary writing, I touch on these varied pieces to elucidate the rise of "systems thinking" and "electrical utopianism" in *fin de siècle* American culture. I also draw these disparate texts into conversation with Gilman in order to honor her interest in systems and her emphasis on interdependence. In view of her lifelong battle against the idealization of individualism, Gilman's legacy demands that we also contemplate the connections that energized her work and contributed to its meaning. This chapter, then, traces Gilman's intricate depictions of systems across her nonfiction and then her fiction, pausing to reflect on the resonances and dissonances of other utopianisms that developed in feedback with her writing and with the expanding electrical systems that consumed the attention of many great, hopeful minds.

THE HUMAN STORAGE BATTERY

In her 1898 treatise *Women and Economics*, Gilman briefly introduced a metaphor that she would continue to develop throughout her career: the human storage battery. This early sociological treatise mentions the "vast storage battery of female energy" only in passing (Gilman 1998, 68). In *Concerning the Children* Gilman develops this image further: "The human creature does not originate nervous energy; but he does secrete it, so to speak, from the impact of natural forces. He has a storage battery of power we call the will" (1901, 47). By the time she published *Human Work*, Gilman had developed a cosmology that described human relationships as "held together in definite relation by laws of attraction and repulsion" (1904, 112). She had re-inscribed social interactions of men and women as a natural law akin to the laws of electricity and magnetism.[6]

Gilman's narrative strategy was not groundbreaking. Professionals in sociological fields commonly derived concepts from the sciences to garner the cultural credibility of these disciplines. This fact remains imprinted upon that ubiquitous moniker, "social scientist." Of course, practitioners didn't always understand themselves to be borrowing from these traditions. The sociologist Lester Frank Ward (recognized by Gilman as an influence in her adopted field) insists that human desire is no different from other physical forces: "The central and all-important truth toward which all that has been said thus far in this work has tended, is that *desire* is a true natural force. There is not the least, figurativeness, metaphor, or analogy in this formula. It is the expression of a literal truth" (1893, 94). Though in hindsight we may argue that Ward's defensiveness calls attention to his inheritance from Newtonian physics (a field he mentions on the same page), it is worth remembering that social scientists such as Ward actually understood themselves to be uncovering truth in the same way that experts in "hard" scientific fields fancied themselves to be doing.[7]

Gilman inherited Ward's inclination for scientific appeal. Though she may have been less likely to dismiss the power of metaphor or analogy, she shared with Ward a certainty that her treatises told the truth. She even prefaced some of her fiction works with citations and allusions to scientific literature, casting her imaginative literature as a usable extension of sociological or medical research.[8] Gilman's scientism stands apart from the scientism of some of her contemporaries not because it is novel, but because it is inconsistent. As I contended in my introduction, the hybridity that might call into question the "literary value" of her work renders her rhetorical play all the more interesting.

Now let us reconsider the passage about "attraction and repulsion" that I quoted above in its broader context. This passage begins with a strawman argument. The silliness of Gilman's analogy to frogs' eggs allows her subsequent claims to seem all the more reasonable:

> Human beings are not webbed together like frogs' eggs, but they are held together in definite relation by laws of attraction and repulsion, like the constituents of any other material body. The stuff that Society is made of is thickest in great cities, and as it develops these dense and throbbing social ganglia grow and grow. In wide, rural areas the stuff is thin—very thin. But watch the lines of connection form and grow, ever thicker and faster as the Society progresses. The trail, the path, the road, the rail road, the telegraph wire, the trolley car. ...
>
> The social organism does not walk about on legs. It spreads and flows over the surface of the earth, its members walking in apparent freedom, yet bound indissolubly together and thrilling in response to social stimulus and impulse. (Gilman 1904, 111–112)

This excerpt exemplifies another of Gilman's signature narrative moves. Its description of physical systems—transportation and communication systems in this example—leads seamlessly into a discussion of invisible, figurative interconnections. Much as in the article "Touching the Button" (discussed in chapter 2 above), Gilman hypostatizes social connections among individuals by equating them to actual cables, tracks, and wires. Yet where "Touching the Button" draws an analogy between human decisions and electric buttons, it is difficult to classify Gilman's language as strictly analogical. She blends biological and electrical imagery to describe the "stuff that Society is made of" as both tangible and intangible, both animate organism and inanimate wire.[9]

To unpack the meaning that Gilman builds into her biological-and-electrical imagery, reflect first upon the provenance of this image in poetry: Whitman's "I Sing the Body Electric," which opens with a brief mention of interconnectivity: "They will not let me off till I go with them, respond to them, / And discorrupt them, and charge them full with the charge of the soul" (Whitman 1921, 81). After he transmits his "charge" to others, Whitman begins cataloguing their bodies. Itemizing the fleshy and the spiritual parts of all humans,[10] Whitman insists that all, regardless of their hierarchical valuation in American society, are made up of "meat" and "soul." In other words, he shifts his focus from the distribution of energies among bodies to the inherent interchangeability of all individuals.

Gilman never explicitly likens her human storage battery to Whitman's "body electric," but she does quote his "Song of Myself" in *Human Work*. Although she famously extolled the virtues of Whitman's poetry elsewhere, in *Human Work* she criticizes his fixation on the "contented individual animal" (Gilman 1904, 102).[11] Lamenting that Whitman tantalizes readers with an alluring but unattainable "invitation to give it all up and go back to the beginning," she urges readers to undertake the difficult but rewarding task of re-imagining themselves on a larger, systemic scale (ibid., 103). This appraisal of Whitman situates Gilman's human storage battery as, among other things, an updated rendition of his work. In this context, Gilman can be said to emphasize the interconnectivity that Whitman quickly abandons.

Gilman and Whitman predicate their "bodies electric" on different understandings of electricity, further demonstrating how constructions of the human self developed in feedback with constructions of electrical systems. When Whitman composed "Song of Myself," electricity was predominately accessed through voltaic piles and static charges. By the time Gilman incorporated his imagery into her "human storage battery," this energy was more often distributed via systemic connections from distant generators. Gilman thus revises Whitman's "body electric" by aestheticizing this new emphasis on interconnection. She describes each person as a "storage battery of nerve force," but she insists that none radiates energy alone (1904, 19). Each must be put in circuit with other batteries to survive. She explains: "Our connection is so subtle, so fluent, each human brain being so large a storage battery of social energy, that we can separate for a time with no loss. But make the separation complete and the humanness dies" (114). Translating the noun *human* into the adjective *humanness*, Gilman posits a proportional relationship between connectivity and humanity. According to this eugenicist ratio, each person can become wholly human only when linked with others in an "interminable array of batteries, full charged" (388).

Extending the metaphor, Gilman describes limited interactions, such as those between two men working in a lighthouse, as "a dangerously 'short circuit'" (92). This passage plays with the newly ambiguous connotations of the word *short*. In the parlance of electrical experts, a short is a disruption in the flow of energy across a circuit of any size. Gilman transforms this technical term into a pun, simultaneously calling up the conventional connotations of the word when she claims that more effective circuits would be long as opposed to short. She uses the slipperiness of this new vocabulary to suggest that an optimally efficient circuit would link a large number of human batteries to one another.

Such depictions of "short" circuits and failed connections recur throughout *Human Work*. Later in that treatise, Gilman describes the inability of individuals to exchange energy as the only significant impediment to social progress. She argues: "With our vastly increased capacity for happiness our misery must be accounted for by 'failure to connect' with the universal energy in one or both ways. We are denied our share of stimulus, we lack social nourishment, or, worse, we are denied our right to discharge, are not rightly placed in the field of social action, are not doing the work which belongs to us" (205). Attributing "misery" to an interrupted connection, this passage re-formulates the famous critique of the "rest cure" that Gilman memorialized in "The Yellow Wall-Paper."[12] Whereas Gilman's early short story personalizes the plight of a woman who is forced by her doctor and husband into maddening bed rest, this treatise generalizes the problems that result from isolation. It asserts that all human life demands interconnection in order to frame seclusion as unnatural and injurious.

Passages like the one quoted in the preceding paragraph proliferate in Gilman's body of work, revealing how she agglomerates different discourses to promote systems thinking. In the aforementioned passage, "failure to connect" and "universal energy" connote electrical circuits and physical laws, while "denied our right to discharge" frames these scientific constructs in legal or ethical terms. Gilman blends organic and electric metaphors and couches both in the political terms of early-twentieth-century feminist thought. Her pastiched narrative raises interesting questions about the spreading of metaphors across different disciplines, but it also threatens to undermine Gilman's argument. Contrary to her stated intentions, she risks making her natural laws appear socially constructed rather than organic by describing them as "rights."[13]

All three frames of reference that Gilman uses—legal, biological, and electrical—posit the social change she promotes as an *a priori* condition that has been unduly interrupted. Rights have been denied. Energy exchanges have been suspended. Nourishment has been withheld. This narrative strategy underscores the urgency and the naturalness of interconnection. It sounds a chord that reverberates throughout Gilman's body of work as she argues that fantasies of isolation and individuation disproportionately harm women. From her short story "The Yellow Wall-Paper" through her sociological treatises and her utopian fictions, Gilman cultivates empathy for the women who are shut up in their houses and disconnected from the wider world. To advocate the inclusion of these alienated women in social life, Gilman criticizes the cultural logic of individualism, arguing that interconnectivity is a physical law, a biological imperative, and a technical condition of modernity. In *Human Work*, she quibbles with the notion that the "individual" can even exist. She anticipates posthumanist philosophies when she ridicules Americans for failing to address the "social organism" because they have been duped by the "temporary detachableness of the individual human being" (Gilman 1904, 113).[14]

Gilman's depictions of electrified bodies make up one compelling set of metaphors she uses to motivate these broader arguments about interconnectivity. Among her varied descriptions of electricity, her image of the human battery is particularly evocative because it is nonspecific. It eschews gender, race, and class markers, suggesting that all people are equally in need of connection. This metaphor revises prevailing constructions of differentially gendered "bodies electric." As the scholar of American Studies Carolyn Thomas de la Peña argues (2003, 125), representations of men and women's relationship to electricity differed during Gilman's era: "Turn-of-the-century promotions of electric products for men showed the body visibly charged by external forces. ... Women's 'electrified bodies were rendered relaxed, soft, beautiful, and often prone ... they radiated not an external change but an internal glow." Gilman neutralizes this gendered difference, casting masculine and feminine bodies as equally electrified.

While Gilman's electrical icon seems to promote pluralism, she writes her batteries into a hierarchical symbolic system. She consistently defines humanness in terms of social and technical interdependence, stating explicitly in *The Man-Made World; or, Our Androcentric Culture* that a lack of interconnection makes some "less human" than others (Gilman 1911a, 16). This claim devalues rural people, non-white people, and others who didn't regularly use modern systems. In *Human Work*, these implications appear more diffusive. For example, just before she discusses her laws of attraction and repulsion, she once again likens intangible and tangible systems: "All Social evolution is the story of the development of the connective tissues of Society, from language, the great psychic medium, to steel rail and wire, the infinitely multiplying physical medium" (1904, 112). Correlating the presence of "steel rail and wire" with "language" and "social evolution," Gilman hints that communities that use the latest systems are superior to those that do not.

This rhetoric calls up (even if it arguably repurposes) a racist and sexist understanding of system building that was promulgated by the popular press and by celebrities of the electrical industry. For example, in keeping with his longstanding tradition of boasting and grandiosity, Thomas Edison told *Good Housekeeping* that "appliances in the home would literally force the housewife's brain and nervous system to evolve to be the 'equal' of her husband's" (Marshall 1912).[15] Edison was not the only public figure to advance such a claim. As Thomas de la Peña has shown (2003, 2–3), in the early twentieth century many Americans believed that interactions with electricity would bring Victorian bodies into the modern era. Although this extrapolation remains unstated in this passage, Edison and like-minded contemporaries maintained that people who didn't use electrical appliances in the home would be inferior to those who did. Like Gilman, Edison etches this inequality onto human bodies, inscribing women's "brain and nervous systems" with different degrees of humanness based on their electrical consumption.

For divergent reasons, Gilman and Edison both suggest that the very presence of emergent systems could improve a user's life and mind. At the same time, they tautologically use that claim to imply that electrical systems could function as a metric for their users' worth. Edison's argument is misogynistic. He assumes that women will encounter electricity as consumers in the home, although by that time several women had earned degrees in electrical science and engineering, and Hertha Ayrton had already been elected into the Institute of Electrical Engineers.[16] He asserts that inventions (implicitly developed by men) can help women (implicitly belonging to a class that can afford these new items) to evolve. In her treatises, Gilman circumvents these gendered presumptions by avoiding questions of production and consumption altogether. Her account erases the uses of these networks and the decisions that dictate where "steel rail and wire" are strung in order to characterize these systems as the natural "tissue" of an "organic" whole (Gilman 1904, 113).

This distinction between Gilman and Edison's formulations is fundamentally one of scale. Edison focuses on the individual user and her microevolution. He cannot imagine a social configuration other than the traditional home, and so he hawks task-specific inventions that claim to increase the efficiency of middle-class housewives. In the process, he maps his discussion of electric inventions onto existing notions of romantic individualism.[17] Anticipating the public-relations campaign that General Electric would initiate decades later, he asserts that electrical apparatuses will improve the housewife's very being. As Judy Wajcman (1991, 85) and Ruth Schwartz Cowan (1983, 97–99) have shown, these devices actually reshuffled tasks that were once divided among household workers into a housewife's daily labor.

As I will demonstrate below, Gilman shares Edison's interest in shifting the burden of housework from laborers (many of them non-white) onto electrical appliances. (In fact, the convergence between the industry's desire to encourage electrical use and eugenicists' desire to rid the home of non-white workers might explain the widespread adoption of electricity in the home, even though the applications of this energy didn't appreciably improve the housewife's conventional workday.[18]) By reconceiving of the home as a part of a larger system, she imagines a redistribution of labor that doesn't place the entire burden of housework or emotional labor onto individual wives. Gilman writes about these changes from a distant sociological perspective that doesn't differentiate between masculinized inventors and feminized consumers. From this vantage point, she interprets the appearance of certain structures as evidence of a social phase change.[19] In *Human Work*, the emergence of large-scale systems reveals nothing about the genius of Edison and his laboratory collaborators. Instead, the presence of these systems indicates that

American society has evolved into a new and more civilized form of being. Gilman's assumptions about humanness and electrical consumption remain prejudiced, but her biases are based on class and race rather than gender.

While Gilman's mixed biological and electrical metaphors were unique for their feminism, her decision to invoke electricity as a symbol for "evolution" had many precedents. The electrical industry often described its own development in resonant terms.[20] Indeed, Charles Steinmetz's account (quoted above) offered a congruent explanation without Gilman's fleshy metaphors. As I will demonstrate in my next chapter, an increasingly consolidated electrical industry attempted to control the way Americans understood and used electricity during the twentieth century. While users and other relevant social groups tried to promote alternative imaginaries of the future, the industry's press agents compressed a flurry of changes into a linear account of progress.

According to the industry's narrative, innovative geniuses from Benjamin Franklin through Thomas Edison incrementally improved electricity.[21] An article in the March 1891 issue of *Electric Power*, "The Electric Light Convention," even coined the phrase "A.F. (after Franklin)" to tell this history, as when it says: "In the year 127 A.F. (after Franklin), Edison divided the electric current and produced miniature suns that might be placed wherever light was desired [or used for] pleasure." The story went something like this: In Benjamin Franklin's day, this energy source was generated locally in Leyden jars or with electrostatic machines. After a hundred years of little commercial progress, a burgeoning industry harnessed electricity for communication by constructing extensive networks of electro-magnetic telegraphs and telephones. The pace seemed to quicken from there. In the 1860s and the 1870s, arc lights began to replace gaslights. In 1882, Edison unveiled his direct-current incandescent lighting system that distributed electricity from a central station. A few years later, George Westinghouse Jr. and other industry moguls developed the long-distance transmission systems that eventually traversed the nation. These developments conjoined electricity and interconnectivity in the American cultural imagination.

Accentuating the quickening pace of development, these promotional accounts equated electrification with social progress. They also began to use the word *electricity* as a subject of sentences, so that *electrical progress* came to seem autonomous—detached from the human decisions that structured how electricity was harnessed and used. This decision was financially and rhetorically economical. It allowed promoters to cultivate public interest in electricity itself, benefitting everyone in the industry.[22] At the same time, it enabled them to describe electrical systems as constantly evolving without having to identify explicitly the experts and users whose decisions incrementally inflected the course of electrical development. By collapsing myriad technical choices into a mythologized, linear account of improvement, these narratives also paved the way for technologically fallacious thinking.

Gilman draws on these widely circulated narratives as she elaborates upon Whitman's body electric. However, she doesn't defer to the authority of these expert discourses as Lester Frank Ward does. By frequently focusing on the storage battery, she inadvertently draws attention to the omission of this imperfect artifact from promotional narratives about electrical progress. As readers who were familiar with these devices would know, actual storage batteries lose much of their charge when left disconnected for too long.

Steinmetz helpfully summarizes the ineffectuality of the storage battery in the Camp Co-operation speech I mentioned above. He suggests that experts had made only limited progress in trying to improve them for practical use. He classifies the storage battery as the only serious limitation the electrical industry still faces in the early twentieth century:

> These then are the three advantages—the ease of transmission, the simplicity of conversion into other forms of energy, and the possibility of very high concentration.
>
> Against this, however, stands as the one serious disadvantage of electrical energy, the fact that it can not be stored but must be consumed at the rate at which it is produced. We have, indeed, the so-called electric storage battery, but this does not really store electrical energy, but stores energy by converting it into chemical energy, which by the discharge is reconverted into electrical energy. It is economically very inefficient, and therefore does not come into considerate where we consider and deal with electrical energy in the energy supply of the industries. (Steinmetz 1913, 58–59).

This constraint doesn't concern Steinmetz. It may even bolster his claim that electrical power distribution tended more toward socialism than previous forms of energy consumption. If electricity cannot be taken into isolated homes and stored like coal or wood—if users could best access this power through a system, distributed (supposedly with equal access) to everyone else—then we might see how Steinmetz could interpret electrical power systems as challenging earlier, individualistic uses of energy.[23]

Although capitalist electrical experts didn't share Steinmetz's politics, they did share his assessment of batteries. Technical reports written for popular audiences agreed with Steinmetz's perspective more often than not. For example, the article "Storing Electrical Heat" in the April 1910 installment of *Popular Electricity in Plain English* offers another instructive glimpse of what the storage battery signified at the time: "So far the storage battery has been the only device for storing up electrical energy (in the form of chemical energy) so that it could be used as required, and the storage battery has not been developed to a stage to make it practicable for the household. As a result the electric light plant must be built with a capacity sufficient to meet the demands of the moment of all the patrons of the company, which demands result in high 'peak loads' at certain periods of the day."

This technical vocabulary was important because it developed in tandem with electrical systems and the unique problems they posed: "peak loads" refers to the highest amount of energy consumption during any given day. Because there was no convenient way to store electrical energy in batteries, managing and optimizing these loads became integral to electrical development.[24] Gilman was inspired by this style of systems thinking, though she may not have considered the technical limitations of storing electricity that the texts quoted above describe.

In Gilman's day, load optimization was an important issue because of the inefficiency of batteries. Nonetheless, the myth of eventual perfectibility that I discussed in chapter 2 also inflected public discourses about these batteries. Faith in the pace of research and development allowed the American public to imagine that electricity might soon be stored efficiently in batteries, just as it imagined that the electric chair might soon execute individuals instantaneously despite its many botched attempts. While experts speaking to lay or technical audiences affirmed the impracticability of the battery, newspapers were inconsistent in their characterization of this device. In an article dated March 1, 1888 and headlined "Fourth; New York," the San Jose *Evening News* described the storage battery as a terrible failure; on November 26 of the same year, the newspaper touted the battery's perfection with a piece titled "Electricity" and subtitled "it May be Stored and Used at Pleasure." An indeterminate icon, the storage battery connotes the allure and the difficulty of holding an electrical charge.

The battery's imperfections need not undermine Gilman's metaphor. Since she deploys this image in treatises that depict women as underutilized sources of energy, her battery analogy might appeal more effectively to readers who understood the limitations of this actual artifact. To those readers, the image of the inefficient battery would summon the loaded connotations of *waste* during this Progressive Era moment, raising questions about the rationality of a "separate spheres" ideology that optimizes the energy of men while needlessly depleting that of women.[25] We could imagine that load optimization experts would scoff at such an irrational management of human energy.

In *The Home* (1903) Gilman amplifies this intonation, arguing that considerable social problems arise from the fact that women are given no outlet for their energies. She argues, "We charge her battery with every stimulating influence during youth; and then we expect her to discharge the swelling current in the same peaceful circuit which contented her great-grandmother! This gives us one of the most agonising spectacles of modern times" (Gilman 1903, 220–221). This passage plays on the notion of obsolescence that attends the concept of progress: if modern Americans would not want to use their great-grandmother's sources of power or light, why would they want to preserve her social customs?

Here, Gilman invokes the misleading but prevalent conceits that older electrical systems were completely displaced by recent advancements and that electrical progress has outpaced social progress. But other understandings of electrical development were apparently available to her; elsewhere she describes electrical and social progress as commingled. In *Human Work*, for example, she argues that life adheres to the law of "conservation of energy" and that transportation, communication, and power transmission systems must confer an evolutionary advantage by enabling people to receive and transmit "energy" more easily.[26] She posits a convoluted relationship between actual and figurative systems, proclaiming: "As an effect of changed conditions our conduct to-day is at the grade required by steam and electric communication" (Gilman 1904, 24). The subject of this sentence appears to be "steam and electric communication," two agents that pair uncomfortably with the verb "require." Although what it would mean for communication systems to require anything remains unclear, Gilman goes on to pronounce that electric and steam systems "make for peace today, for smooth and rapid growth of international agreement." According to her developmentalist logic, these systems appear to be both a cause and an effect of the evolution of human civilization (24).[27] In addition to perpetuating a Social Darwinist hierarchy of primitiveness and civilization, such passages emphasize "international agreement" while distracting attention from national discord. Unmentioned in these pages are the victims of electric execution or the many citizens who were restricted either by law or by prohibitive cost from accessing these supposedly beneficial systems.

Less evident than such omissions are Gilman's intended arguments. Like Dreiser, Gilman strives to develop a strategy of narration that can account for the complex, systemic interactions between humans and their inventions. Consequently, her work has similar shortcomings to Dreiser's. Her syntax obscures agency, leaving readers to wonder if she believed that steam and electricity improved human conduct or that human evolution necessitated the invention of these energy sources. Such awkward sentence structure reveals the difficulty Gilman faced when trying to ascribe the agent that actuates change. Today we might say that she cannot assign a first mover because there was not one: humans build their systems while those systems build us in turn. Gilman does in fact recognize this feedback loop elsewhere in *Human Work*, as when she claims that "man's spiritual nature manifests itself through material things, and grows by means of them" (Gilman 1904, 161). Although it would be anachronistic to call Gilman a co-constructionist, she tackles an issue that philosophers and science and technology studies scholars continue to confront: how best to describe a complicated and constantly changing relationship between "humans" and "systems" when these categories lack fixity.[28]

Gilman confronts this issue by employing diverse (and occasionally incommensurate) lines of reasoning—anything to convince her contemporaries to set aside their presumptions about romantic individualism. At times, she insists that the very notion of the individual is a fiction, as no person can survive in complete isolation. More often, she claims that cultural critics cannot rationally conceive of social problems in individualistic terms. In *Human Work* she ventriloquizes proponents of individualistic self-reliance, asking: "Why is it not better to produce and consume locally, each man for himself, as Tolstoi would have us?" She answers this question by claiming that human evolution demands integration, not isolation: "A healthy, growing, social life constantly re-creates its body as does the physical life, and our American civilisation shows this beyond all others in its rapid adoption of new material forms and processes. The constant demand for easier and swifter mechanism is as natural and healthful in society as it is in a physical body, and physical evolution has moved on that line continually" (Gilman 1904, 173).

Describing "mechanism" as "natural," this passage flies in the face of the conservationist writings of John Muir or Teddy Roosevelt. These cultural critics advocated a return to nature to replenish the bodily energies that drained by modern life; Gilman claims that such "'call of the wild' literature" inspired Americans to long for a mythic, unreachable past instead of solicitously building a desirable future. When she asserts that "easier and swifter mechanism" will lead to improved social conditions, Gilman does more than challenge these naturalists or promote industry narratives of electrical and mechanical progress. She also defies the medical orthodoxy of her day, by arguing that the social ills imputed to overcivilization and industrialization actually result from the American obsession with individualism.

BEYOND BATTERIES

Although Gilman repeatedly points to electricity as a symbol and agent of progress, she doesn't describe every modern system as civilizing or revitalizing. In *The Home*, for example, she depicts the inescapable interconnectedness of the modern domestic space as a drain on women's energies: "The tradesmen, in a city flat, are kept at a pleasing distance by the dumbwaiter and speaking tube; and among rich households everywhere, the telephone is a defence. But, even at such long range, the stillness and peace of the home, the chance to do quiet continued work of any sort, are at the mercy of jarring electric bell or piercing whistle" (Gilman 1903, 44). The interrupted "stillness" and the "jarring" noise of the bell denote an enervating soundscape that appeals to contemporaneous medical discourses. Gilman's formulation of the tired female body here and in her fiction echoes discourses about a new condition called neurasthenia—a condition with which Gilman suffered.[29] In fact, electro-medical practitioner George Miller Beard (who shared a practice with A. D. Rockwell, the electro-execution expert mentioned in chapter 2) coined this diagnostic term to describe a specific kind of depletion that resulted from "overcivilization," the result of enduring stresses akin to those Gilman enumerates.

Beard's account of this condition is instructive here both for its consonances and dissonances with Gilman's. In *American Nervousness*, the 1881 treatise in which he elaborated upon his own definition of neurasthenia, he describes a specific version of the "body electric," using "Edison's electric light" as a model for the body's functions and dysfunctions. I quote Beard at length to illustrate how integral electricity was to his understanding of the human body:

> Edison's electric light is now sufficiently advanced in an experimental direction to give us the best possible illustration of the effects of modern civilization on the nervous system. An electric machine of definite horse-power, situated at some central point, is to supply the electricity needed to run a certain number of lamps—say one thousand, more or less. If an extra number of lamps should be interposed in the circuit, then the power of the engine must be increased; else the light of the lamps would be decreased, or give out. This has been mathematically calculated, so that it is known, *or believed to be known, by those in charge*, just how much increase of horse-power is needed for each increase in the number of lamps. ... *The nervous system of man is the centre of the nerve-force supplying all the organs of the body.* ... When new functions are interposed in the circuit, as modern civilization is constantly requiring us to do, there comes a period, sooner or later, varying in different individuals, and at different times of life, when the amount of force is insufficient to keep all the lamps actively burning ... this is the philosophy of modern nervousness. (Beard 1881, 98–99, emphasis added)

According to Beard, electricity is simultaneously a model for the body, a strain on the body, and a cure for strains on the body. At times, he seems uncertain about the tenability of his comparison, as when he appeals to the vague authority of "those in charge." Still, he teases this analogy out for two full pages, indicating the allure this comparison held for him.

To some extent, this definition of American nervousness promotes Beard's preferred methodology of electro-medicine. By likening the tired body to an electrical system without quite enough juice, Beard hints that patients should turn to his electrical apparatuses for a healing jolt. But just because Beard monetizes this construction of the body doesn't mean that he understood it as any less of a fact. As Laura Otis and Timothy Lenoir have argued (1994, 185–207), Hermann Helmholtz and Emil DuBois-Reymond had understood the body as functionally electrical since 1851. Otis explains that these "comparisons between organic and technological systems were not mere devices for popularization but became incorporated into the scientists' own vision and understanding of the nervous system" (2002, 106). The physiologists of whom Otis and Lenoir speak had modeled the nervous system on the telegraph on the basis of experiments that had wired flesh into circuits with electrical apparatuses. Beard too applied electrical currents to patients' bodies before he proposed this comparison.[30] His use of "Edison's electric light" in *American Nervousness* serves as both an analogy and a truth claim. At the rhetorical level, Beard alters Helmholtz's model slightly to render it more easily discernible: dimming lights of a strained circuit are easier to visualize than diminished energy passing invisibly through a telegraph wire. In his medical practice, Beard also uses the newest electrical equipment available to expand prevailing conceptions of human physiology.

Beard's construction of the body as a special kind of powerhouse was also popular outside of esoteric medical literature.[31] It appeared in advertisements for new healing devices, for exercise apparatuses, and for vacations to the West. It pervaded reports of electric execution—including the article by Howells, mentioned above in chapter 2, in which he wryly refers to the condemned person as both "victim" and "patient" (1888, 23). It was also taken up in research by social scientists. For example, Ward argues: "The psychic force may also be likened to electricity. The animal body may be regarded as a battery which it is the function of life to keep constantly charged. The attachment is to the muscles, and locomotion is analogous to that of an electric car. Again, the nervous system is analogous to a telegraphic or telephonic system, the fibers representing the wires" (1893, 95). This passage combines different aspects of electrical imagery. The body is like a battery because it can lose its charge; the body is like an electric car because it locomotes; the body is like an electrical communication system because its nerves branch rhizomatically, conveying impulses. Ward doesn't find a single appropriate electrical analogue. He couples several together to approximate the electrical quality of the "animal body" that he strives to express.

Gilman was a student of Ward and of Whitman, and her allusions to electrical energy can be said to build upon both poetic and scientific traditions. And build she does. If Beard, Whitman, and Ward all understand the individual body as an electrical system of some sort, Gilman reframes these individual bodies as nodes in a much more expansive electrical network. In her nonfiction, the body is not a microcosm of an electrical system. Rather, she visualizes the electrical system as the nervous tissue that links individual humans (or "cells") into a larger social body. Once again, this shift in scale maps onto a shift in electrical material culture, demonstrating how metaphors and material systems evolved in tandem. When Beard postulated his model of the "body electric," Edison's DC plant could transmit energy only over short distances before losing much of its charge; power stations were relatively local. By the time Gilman adopted this partially scientific and partially poetic trope, electrical power was entering American homes from increasing distances.

I will discuss the rise of long-distance transmission systems in the next chapter; what interests me here is Gilman's claim that persisting social ills follow from Americans' inability to understand themselves from this macrocosmic perspective: "We are treating social disease by local application … but we forget, or do not know, that this local trouble … is on a living body, and is caused and maintained by diseased conditions in that body, far beyond the material boundaries of that location" (1904, 383). Although Gilman's description of the problem remains in medico-political register of the diseased body politic undergoing treatment, her description of the solution relies on connotations of electrical power. Here we finally see the contours of the macro-scale "body electric" she alludes to throughout the treatise:

> To feel the extending light of common consciousness as Society comes alive!—the tingling "I" that reaches wider and wider in every age, that is sweeping through the world to-day like an electric current, that lifts and lights and enlarges the human soul in kindling majesty:
>
> To feel the power! the endless power! Not only the ceaseless stream of the universal Godness, but our interminable array of batteries, full charged; the stored energy of all time embodied in poem and story, in picture and statue, in music and architecture, in every tool, utensil, and giant machine wherein the human brain and the human hand have made force incarnate. (388–389)

Peppered with exclamation points, this passage imagines how sensory experience can extend beyond the boundary of the skin. In it Gilman contends that "Society" can evolve into a compound sensate body—the "tingling 'I.'" Borrowing again from Whitman, this passage re-codes the personal pronoun as a collective entity. Describing "common consciousness" as "tingling ... like an electric current," Gilman depicts social change as an "endless power" that she can "feel"—perhaps in the way that a neuron might figuratively "feel" the electrical charge of an epiphany firing to other neurons and collectively forming an idea.

Throughout *Human Work*, Gilman portrays human energy as precious and fleeting. Her human batteries are in constant need of recharging. In the passage quoted above, she reveals the stakes of nurturing the energy she describes. She avows that humans, when "full charged," can transform their energy into material, social, and moral artifacts. She submits that humans turn energy into culture, by which she means a living repository of human experience. She characterizes art, architecture, literature, and machines as batteries that hold "the stored energy of all time." Unlike that imperfect storage battery (the individual human body), the facts and artifacts of human creativity need never lose their charge. It is as if Gilman anticipated that, decades hence, scholars like me would thrill as we unlock the energy and ideas she and her contemporaries poured into their work.

GILMAN THE UTOPIAN

The exultant rhetoric of *Human Work* marks a utopian turn in Gilman's writing. In her earlier treatises, she discussed electricity and interconnectedness in matter-of-fact terms, deploying scientific analogies to garner cultural credibility. In *Concerning the Children*, she invokes her "storage battery" metaphor only briefly. More often she depicts electricity as one of many modern systems that connect individual homes into a larger social body. In *Human Work* all life throbs, connected electrically, like nerves. Gilman's fictional narratives navigate between these poles. She frequently sets her utopias in romanticized power "grids"[32] that heal ailing bodies, but these fictional narratives appear more realistic and less rhapsodic than *Human Work*. They often describe women managing, rather than constituting, electrical networks. This change in tone follows from a change in conversation: whereas Gilman's nonfiction builds upon work by Lester

Frank Ward and Walt Whitman, her utopian fiction responds to Edward Bellamy. As Polly Wynn Allen argues (1988, 86), "Gilman was an important recruit to the Nationalist cause. She whole-heartedly supported Bellamy's call for state-supported domestic services as part of a sweeping social transformation." Consequently, to understand the intervention Gilman's utopias make, we must first reflect upon how they engage with the Bellamy tradition.

Edward Bellamy's novel *Looking Backward: 2000–1887* was more than a best seller. It was a call to action that many publicly heeded. In Bellamy's imagined future, individuals no longer had to fight their way up the social ladder. In the year 2000 the labor system would be managed with military precision. Bellamy's future workforce is a volunteer army that evolves in constant feedback with the desires of the workers. If a job has proved too strenuous to attract volunteers, government officials tweak the parameters of the position until it appeals to a sufficient number of laborers. In an age before cybernetics and computing, Bellamy's vision of relays, feedback loops, and systems management seemed prophetic. Under the spell cast by imaginative writing, *Looking Backward* transforms the idea of electrical load optimization into labor optimization and mood optimization. Bellamy's future generates "peak loads" of contentment.

Readers found Bellamy's critique of the late-nineteenth-century class struggle convincing. More important, they found his proposed future attainable. His fans formed "Bellamy clubs" or "Nationalist clubs" and sought to change public policy. Most of those clubs had dissolved by 1896. Still, their influence endured, attesting that fiction could incite change.[33] The wild success of Bellamy's novel also inspired activists to take to their typewriters. Among the hundred-plus utopias that were published within twenty years of Bellamy's, many responded to his volume by name. Some sought to fine-tune his vision of the future; others aspired to attain similar literary and political prominence. In either case, Bellamy instigated a boom in utopia writing. This fad requires investigation in itself. As Frederic Jameson argues (2005), the utopian genre doesn't appear in every culture during every historical moment. Gilman's utopias can expand our understanding of a unique moment in American literary history while lending further insight into her figurative engagements with electricity.

Like the scores of other writers who responded to *Looking Backward*, Gilman cites Bellamy explicitly. The preface to her first utopian novel, *Moving the Mountain*, claims that the hope for utopian possibilities is a universal human trait, exemplified by the "best known" works of "Sir Thomas More and the great modern instance, 'Looking Backward'" (Gilman 1911b, 5). If the utopian urge is universal, Gilman argues, a better society need not be as "remote" or far off as More or Bellamy imagined. Insisting on her utopia's practicability, she suggests that her novel "indicates what people might do, real people, now living, in thirty years—if they would" (6). Gilman takes Bellamy's vision a step further by suggesting that her readers need not wait a hundred years for social change.

Moving the Mountain also appropriates the central narrative strategy of *Looking Backward*. It too features an outsider narrator, named John, who can compare *fin de siècle* culture against the novel's invented future. As Gilman's novel begins, John's sister, Nellie, finds her long-lost brother in Tibet, where he has suffered a head injury. That plot device allows Gilman to frame her novel as more realistic than its predecessor by modifying the trope of suspended animation. Bellamy's protagonist, Julian West, awakens after a hundred years of mesmeric sleep, while Gilman's John awakens after suffering from amnesia for thirty years. Gilman deploys this trope of "waking up" in overlapping ways throughout the novel. John "wakes up" from his mental haze, providing an occasion for the initiated to explain the social changes that happened in his absence (1911b, 10). And Nellie repeatedly describes the novel's social revolution as a different kind of awakening. Confused by this oversimplification, John finally asks his sister to "tell me how, when and why the women woke up." He pries from his sister a vague answer: "They saw their duty and they did it" (75, 78).

Gilman's central metaphor revises dominant notions of revolution. She uses the everyday, affirmative experience of "waking up" to characterize her social vision as natural and nonviolent, though whispers of violence remain detectible in the eugenicist policing that structures her utopia. Nonetheless, Gilman's envisioned revolution is not the bloody battle Twain's Yankee feared and then incited; it is not the war that seized Boston in *Looking Backward* while Bellamy's narrator slept; it is not affected by the invention of a new and dramatic device. No, Gilman's utopia emerges from the thoughtful repurposing of already available materials. This logic closely parallels her "human storage battery" imagery. In both cases she frames women as a dormant energy source, healthfully unleashed. And in *Moving the Mountain*, as in her treatises, women appear stronger when their energies are harnessed. John quickly notices that his sister Nellie "doesn't look *old*—not at all. Women of forty in our region, were *old women*, and Nellie's near fifty!" (19). This exclamation simultaneously derides Gilman's current moment for the toll it takes on women's bodies, and it codes this new world order as salutary in comparison. In this possible future, women who once fought aging with bed rest, patent medicines, and cosmetic creams find that the best cure for stress is the "collective action" proposed in *Concerning the Children*.

In *Moving the Mountain*, the revitalization of American society (and of women's bodies) involves the widespread repurposing of electrical power, which John first notices as the absence of smoke: "You've ended the smoke nuisance, I'm glad to see. Has steam gone, too?" (58). Nellie's reply introduces electricity to the narrative as evidence of the new society's awareness of public health and safety issues:

> "We use electricity altogether in all the cities now," she said. "It occurred to us that to pipe a leaking death into every bedroom; to thread the city with poison, fire and explosion, was foolish."
>
> "'Defective wiring' used to cause both death and conflagration, didn't it?"
>
> "It did," she admitted, "but it is not 'defective' any more." (58)

Drawing attention to the potential dangers of Gilded Age electrical networks, this passage associates the improvement of electrical systems with public health and safety. It also marks the utopian society as scientifically advanced, lending credibility to the social changes that come with this technical improvement. In this respect, *Moving the Mountain* imagines a future in which women productively take control of all civic affairs, including, in this case, overseeing electrical networks.[34]

Gilman deepens this correlation between electricity and awakening with her figurative language. John asks "what happened to the Four Hundred—the F.F.V's [First Families of Virginia]—and the rest of the aristocracy?" Nellie explains: "The same thing that happened to all of us. They were only people, you see. Their atrophied social consciousness was electrified with new thoughts and feelings. They woke up too, most of them" (1911b, 138–139). This passage likens social enlightenment to electro-medical treatments that stimulate "atrophied" muscles. This diction calls up Beard's construction of the human body as a light-and-power plant that can be radiant or overstressed to the point of dimness. Indeed, Gilman draws upon a similar understanding of human health and illness throughout the novel. As in her treatises, she extends Beard's understanding of the individual body to account for the entire social system. Instead of treating each human as an isolated powerhouse in need of a personalized electrical treatment, Gilman imagines how society might optimize the flows of human energy. Although *Moving the Mountain* lacks the "tingling 'I'" of *Human Work*, this novel understands the relationships between bodies and electricity in similar terms.

Gilman does more than compare humans to powerhouses in this novel—she treats human and electric power generation as one integrated system. The next time John recognizes electricity (once again after recognizing a conspicuous absence, this time of gasoline), Nellie's husband Owen explains that human energy contributes to their society's abundant electrical power:

> "Electric power there too?" I suggested.
> Owen nodded again. "Everywhere," he said. "We store electricity all the time with wind-mills, water-mills, tide-mills, solar engines—even hand power."
> "What!"
> "I mean it," he said. "There are all kinds of storage batteries now. Huge ones for mills, little ones for houses; and there are ever so many people whose work does not give them bodily exercise, and who do not care much for games. So we have both hand and foot attachments; and a vigorous man, or woman—or child, for that matter, can work away for half an hour, and have the pleasant feeling that the power used will heat the house or run the motor." (102)

In Gilman's earlier works, the storage battery signified the imperfections of human energy. In contrast, *Moving the Mountain* challenges Beard's medical program by suggesting that—in a better-managed world—human bodies might lend their energy to electrical systems, rather than relying on electrical inventions to supplement their own depleted power. This passage also demonstrates a shift in Gilman's thinking about electricity as an abstract source of (and symbol for) power. Above I set her treatises against those of John Muir and Teddy Roosevelt, but her utopias complicate this binarism. While Gilman criticizes the individualism Muir and Roosevelt espouse, she expresses convergent concerns for the environment in her utopias. She modifies their aspiration for preservation—a word that denotes stasis—advocating instead a form of electrical-and-social development that would replenish—a word that better captures her progressivist and evolutionary values.

Gilman develops this environmentalist theme in the utopias she published shortly after *Moving the Mountain*. Her 1913 story "Bee Wise" elaborates upon the novel's depictions of revitalized bodies and communities, but it occurs in the present and imagines change on a smaller, more accessible scale. The narrative depicts a group of enterprising women who develop two towns, one called Beewise and one called Herways. Each is a "rational paradise," "free of the diseases of cities" (Gilman 1999, 271). Tucked into the California coastline, the towns later become franchiseable models that "paved the way for so many other regenerated towns" (267). "Bee Wise" draws out the importance of collaborative specialization, depicting how an assemblage of experts can create an efficient and healthy social system.

The story begins when a woman known affectionately as "the Manager" receives ten million dollars and two large plots of California land from an elderly uncle. Gilman describes this inheritance as a rational (and eugenicist) decision rather than a chance occurrence, conjuring but complicating the trope of the wealthy benefactor from the conventional rags-to-riches story line. The Manager explains that her great uncle "hired people to look up the family and see what they were like—said he didn't propose to ruin any feeble-minded people with all that money. He was pleased to like my record" (266). The Manager's great uncle was a prospector in the Old West, and with this transaction the narrator insinuates that intelligent women are the rational successors of the frontier legacy. The Manager uses this windfall to set up a model town, "a place of woman's work and world-work too" that will "set a new example to the world" (266). It is to be a secular, feminist "city on a hill."

In "Bee Wise," electricity represents more than an incidental energy choice or a symbol of modernity: it is the model on which the entire town is developed. As in *Moving the Mountain*, the generation of electric power becomes central to the larger goal of proving "that a group of human beings could live together in such wise as to decrease the hours of labor, increase the value of the product, ensure health, peace, and prosperity, and multiply human happiness beyond measure" (271). She adds:

> The first cash outlay of the Manager, after starting the cable line from beach to hill which made the whole growth possible, was to build a reservoir at either end, one of which furnished drinking water and irrigation in the long summer, the other a swimming pool and steady stream of power. The powerhouse in the canon was supplemented by wind-mills on the heights and tide-mill on the beach, and among them they furnished light, heat, and power—clean, economical electric energy. Later they set up a solar engine which furnished additional force, to minimize labor and add to their producing capacity. (267)

Typical of early-twentieth-century literature of the American West, "Bee Wise" links power generation with irrigation and water access. Gilman includes leisure in her adaptation of this model, adding a swimming pool to the usual array of "drinking water and irrigation ... and a steady stream of power." According to this passage, most of the labor of establishing Beewise and Herways involves exploiting every available natural power source to maintain a steady flow of "clean, economical electrical energy." Here Gilman obscures the labor of building and maintaining power-distribution systems to portray electrical energy as inherently clean and generative, much as she did in her nonfiction.

In the aforementioned 1912 *Good Housekeeping* interview, Thomas Edison uses a vocabulary almost identical to Gilman's to claim that electricity could free up "mental energy" and thereby revitalize womankind: "To diminish the necessity for utilizing man himself, or woman herself, as the motor-furnishing force for this life's mechanical tasks, is to increase the potentiality of humanity's brain power. When all our mental energy can be devoted to the highest tasks of which it may be capable, then shall we have made the greatest forward step in this world's history. ... It is there that electricity will play its greatest part in the development of womankind. It will not only permit women to more generally exercise their mental force, but it will compel this exercise" (Marshall 1912). Like Gilman, Edison purports that certain applications of electric power will release women from psychically draining domestic tasks that inhibit their ability to participate in more abstract or "higher-order" levels of thinking.

Edison suggestively depicts electricity as a new "handmaiden" that helps women with the domestic tasks they will continue to perform in their individual homes—a feminized version of the electrical servant. In contrast to Edison who describes women as consumers, Gilman depicts women and men as potential managers of power plants. Her townspeople challenge gender stereotypes by allowing women to choose their preferred type of labor—even if that means regulating the electrical engines: "the men ... built and dug and ran the engines, the women who spun and wove and worked among the flowers, or vice versa if they chose" (Gilman 1999, 270).

The stakes of this nonchalant "vice versa" become clearer alongside prevailing notions about women and electrical expertise. Ruth Oldenziel (1999, 16) identifies the 1890s as a moment in which a "white, middle-class, and gendered male engineering identity was shaped in competition with female professional models." Gilman's story appears twenty years later, after this gendering of technical expertise had gained prominence. "Bee Wise" obliquely gives a voice to the women electrical experts who were systematically silenced by the masculinization of their profession. This detail demonstrates a subtle distinction from Gilman's treatises. In her nonfiction, she avoids questions about the gender of producers and consumers; in this short story, she challenges directly Edison's insinuation that women can only occupy the latter, passive role.

More radical still, Gilman uses the image of the power system to render the home-based social unit obsolete. In "Beewise" and "Herways," "There were no servants in the old sense. The dainty houses had no kitchen, only the small electric outfit where those who would might prepare coffee and the like. Food was prepared in clean wide laboratories" (Gilman 1999, 270). Gilman supplies her characters with a "small electric outfit" so that they may serve their own whims. This detail implies that the redistribution of domestic labor need not impose upon individual desires. Still, in these fictional towns food is generally produced and distributed from a central station. Modeled after and integrated into electrical systems, the kitchenless home represents the pinnacle of Gilman's feminist innovation.

By innovatively removing the kitchen from the individual home, Gilman imagines a new way to align home and social life. Yet her "architectural feminism," as Polly Wyn Allen (1988) calls it, doesn't improve the lives of all women equally. This story and Edison's interview both imagine solving the "labor problem" with electrical interventions, and several flaws in their logic become apparent in hindsight. Today we know that "labor-saving" devices ironically increased women's labor (Cowan 1983). More troubling, these narratives conspicuously omit any discussion of how these changes affected those "servants in the old sense." Even the phrase "servants in the old sense" casts the people who filled these roles as obsolete tools. Objectified, they appear to be a kind of force that even efficiency enthusiasts of the Progressive Era were inclined to waste.

Gilman and Edison do not address these issues. Instead, they invite their readers to notice the trappings of "progress" and to ignore other social concerns. By this same reasoning, the designers of the Chicago's 1893 World Columbian Exposition encouraged attendants to marvel at new inventions and to overlook those whose struggles and successes were *not* represented therein—unless they happened to meet Ida Wells-Barnett in front of the World's Fair gates, handing out her co-authored pamphlet "The Reason Why the Colored American Is Not in the World's Columbian Exposition" (1893). At the turn of the twentieth century, writers, architects, event planners, and electrical experts all used electricity to create spectacles of "progress" that distracted national attention from the many ways the country had socially stagnated or regressed.

Although Gilman was an outspoken eugenicist, the examples I have considered sidestep these social issues by manipulating the reader's frame of reference. In her treatises, Gilman avoids questions of racial or gender bias by speaking from an impersonally distant perspective. Although she calls some peoples "less human" than others, she cloaks her racism in the rhetoric of scientific progress that also legitimized gender hierarchies.[35] She obscures the work of system builders, never asking whether power lines were intentionally constructed to connect some people and to exclude others. From this perspective, systems connote progress and evolution. In her fiction, Gilman takes an inverse approach. She focuses closely on white middle-class women, who, she contends, have a right to tap into electric energy. Her fiction purports to save white women from loneliness and drudgery, while also implicitly saving them from relying on the labor of lower-class minority women.

Gilman's racism subtly underwrites the assumptions that shape her utopias. She consistently fantasizes about controlling the composition of her societies. Indeed, the notion of intentional community seems to have been so attractive to Gilman that her short story "Dr. Clair's Place" depicts a selective sanitarium as an ideal home. This 1915 story expands upon themes of her more widely read story "The Yellow Wall-Paper." Whereas the earlier piece detailed the failures of a dominant treatment for neurasthenia, "Dr. Clair's Place" describes an effective cure—one that narrativizes Gilman's understanding of the body as a battery.

"Dr. Clair's Place" begins by introducing a depressed protagonist, Octavia Welch. Readers first meet her on a train, where she explains her resolution to commit suicide: "'Health—utterly broken and gone since I was twenty-four. Youth gone too—I am thirty-eight. Beauty—I never had it. Happiness—buried in shame and bitterness these fourteen years. Motherhood—had and lost. Usefulness—I am too weak even to support myself. I have no money. I have no friends. I have no friends. … I have no hope in life.' Then a dim glow of resolution flickered in those dull eyes. 'And what is more I don't propose to bear it much longer.'" (Gilman 1992, 318) The "dim glow" evokes Beard's image of the overtaxed electrical system. Octavia is a powerhouse undergoing a "brown-out." After Octavia delivers this monologue, the narrator reveals that she has shared this intimate confession in a strange context: "It is astonishing what people will say to strangers on the cars." The anonymity of this interaction underscores the protagonist's desperation. Octavia embodies the disrupted mental energy Gilman generalizes in her nonfiction. The neurasthenic protagonist's body and mind have been wrecked by modern American life. On the train, in the city, the stranger she confides in offers nothing but "cheerful commonplaces" (318). Thus, the short story opens by establishing the lack of sympathy or of resources available for women before it posits a utopian solution.

Although Octavia's story doesn't impress the stranger in whom she confides, another passenger—an old patient of Dr. Clair's and the story's unnamed narrator—serendipitously overhears her. The narrator suggests that—even though Octavia lacks heath, youth, beauty, happiness, motherhood, usefulness, money, or friends—her life might be valuable to science: "If you had an obscure and important physical disease you'd be glad to leave your body to be of service to science, wouldn't you? … You can't leave your mind for an autopsy very well, but there's one thing you can do—if you will; and that is, give this clear and prolonged self-study you have made, to a doctor I know who is profoundly interested in neurasthenia—melancholia—all that kind of thing. I really think you'd be a valuable—what shall I say—exhibit." The stranger doesn't address Octavia with sympathy. Instead she suggests that she can spend the final days of her life "purely as a scientific experiment," adding "There are others who may profit by it, you see" (319). The narrator dispassionately suggests that passively contributing to science might give this depleted woman a new purpose.

When Octavia arrives at "the Hills," she finds that Dr. Clair talks "not as a physician to a patient, but as an inquiring scientific searcher for valuable truths" (Gilman 1992, 324). This passage provocatively implies that a "physician" has a more fraught relationship to a patient than a "scientific searcher" has to her object of study. Dr. Clair, introducing herself, says: "Please understand—I do not undertake to cure you; I do not criticize in the least your purpose to leave an unbearable world. That I think is the last human right—to cut short unbearable and useless pain. But if you are willing to let me study you awhile and experiment on you a little—it won't hurt, I assure you" (323). The doctor begins her search for "valuable truths" by drugging Octavia, forcing the "experiment's" weary body to rest. The premise of this short story, then, hinges on the ethical appeal of scientific experimenters. In Gilman's day, this trust in the medical establishment itself is racialized as white. For example, black women and men were not given the option of informed consent even after this doctrine was formalized with the Declaration of Helsinki.[36]

Interestingly, Octavia's race is ambiguous in "Dr. Clair's Place." If we read Octavia as a white character (as seems likely in view of her ability to move across the country without fear), then this character's willingness to submit to Dr. Clair can seem utopic. To a privileged set of readers, Gilman implies that a feminist approach to medical science could improve lives by fostering cures that other doctors have been unable to find. In this reading, Gilman anticipates the strategies of feminist science studies. If, however, we interpret Octavia as a non-white character, this story would seem to imply that non-white subjects elected to submit to medical experimentation. The ambiguity of Octavia's physical descriptions affords readers the opportunity to read this story as a utopia or as a perverse rationalization for the history of racist medico-scientific practices.

These racial dynamics remain unspoken in the story, as Dr. Clair revitalizes Octavia by instructing her to control electrical circuits (including a colored electric light display and "a light moveable telephone" that pipes various kind of music into the room) according to her own whims (Gilman 1992, 324–325). Like Twain's novel *A Connecticut Yankee* and Edison's *Good Housekeeping* interview, "Dr. Clair's Place" hints that the experience of manipulating an electric control panel could amplify the user's sense of control over his or her own body. By managing light and sound through these circuits, Octavia strengthens her body and mind. However, it is worth noting that the treatment which cures Octavia is not electrical; it is "the right Contact, Soul to Soul." In "Dr. Clair's Place" electricity plays an important supporting role in the healing process, but it is not an end in itself. Throughout this story and similar utopias, Gilman models a different approach to neurasthenia than the electrical cures that George Miller Beard promotes. Whereas Beard treats his patients by supplementing their depleted energy, Gilman invokes electricity to imagine alternative ways of organizing space and, concomitantly, social relations in space. Octavia did not need electrical stimulus to recover but, like a human storage battery, she did need interconnection.

UNDERSTANDING ELECTRICAL UTOPIANISM

Commensurate images pervade Gilman's utopias, extending the "body electric" imagery from her treatises. In her most famous utopia, *Herland*, women control "electric motors" and a wearied protagonist feels energy radiate from his core to his extremities, much like a central station would distribute its energy. "Aunt Mary's Pie Plant," her 1908 short story in *Women's Home Companion*, similarly operationalizes electrical power as a referent and a metaphor. Gilman doesn't depict electricity as a panacea in these stories, but she does draw on tropes from technological utopias or electrical utopias to portray electricity as healthful and helpful.[37]

Why did Gilman repeatedly return to fantasies about electrification in her feminist utopias? Neil Harris (1990), Kenneth Roemer (1976), and Howard P. Segal (2005) might answer this question by exploring how Gilman's utopian representations of electricity respond to fears about her current moment. Indeed, Gilman weds her hopes for the second industrial revolution[38] to her concerns about the first. For example, in *Moving the Mountain,* she makes conspicuous the failures of mechanization even as she expresses her hopes for electrification: "[E]lectric power has removed the worst evils. There is no smoke, dust, cinders, and a yearly saving of millions in forest first on the side! Also very little noise. Come and see the way it works now" (Gilman 1911b, 71). Here Gilman conjures electrical industry discourses about the benefits of electrification by erasing the "smoke, dust, cinders," and noise that accompanied the rise of the machine.

The inextricability of these fears about the current moment and fantasies about the future bring Dreiser's cynicism and Gilman's utopianism into closer alignment than I indicated at the beginning of this chapter. For generations, the cultural meanings of electricity—and more recently of technology—evolved from a dialectic tension between these pessimistic and affirmative interpretations: our advertisements continue to depict cutting-edge devices as nearly magical, and our fictions continue to explore the psychological trauma of the alienated individual who navigates complicated systems, aware that her desires can never really be satisfied by these technologies.

Still, recognizing this dialectic of hope and fear doesn't explain why utopian fantasies crystallized around specific images. The question remains: Why would Gilman and many other utopian writers put their faith in electricity, specifically, after previous inventions had failed to deliver upon similar promises? This question has perplexed a range of scholars—especially because this utopianism was expressed by experts, as well as lay writers such as Gilman. For two hundred years, a diverse group of electrical experts, politicians, scientists, and writers have consistently imagined that—despite all precedents to the contrary—electricity would solve social ills. James W. Carey and John J. Quirk address this conundrum in their 1970 essay "The Mythos of the Electronic Revolution," a study that remains remarkably resonant today. That two-part essay, published in *American Scholar,* delineates the astonishing parallels between nineteenth-century faith in electricity and twentieth-century faith in electronics. Of the late nineteenth century, Carey and Quirk (1970, 228) write that "electricity promised, so it seemed, the same freedom, decentralization, ecological harmony and democratic community that had hitherto been guaranteed but left undelivered by mechanization."

Thomas Parke Hughes also tackles this oddly resilient utopianism in his 1988 article "The Industrial Revolution that Never Came," attributing it to optimism about early-twentieth-century public policies. Of particular interest to Hughes is the failed public electrification initiative called "Giant Power." Proposed in 1925 by Governor Gifford Pinchot, the Giant Power project intended to electrify Pennsylvania in order to foster a wide-ranging social transformation. Pinchot imagined that interconnecting power systems would create robust decentralized networks. Using utopian language, Pinchot projected that interconnected electrical systems would improve the lives of housewives and of all citizens. "In his optimism," Hughes writes, "Pinchot also predicted that electric power would reverse the trend of industrial concentration, mass factory labor, and noisome slum cities."[39]

In view of the fact that technical experts, politicians, and laypeople shared similar hopes for the transformative power of electrical systems, this prominent strand of utopianism cannot be dismissed as naive escapism. To comprehend it, we must recognize that Pinchot, Gilman, and other reformists only imagined that electricity could realize its progressive potential if controlled by the "right" people. To Pinchot electrification would be socially beneficial only if publicly owned; if controlled by private industry, it would worsen corporate greed. Gilman similarly interpreted electrification as progressive only when managed by men and women—specifically those who shared her social values.

While control was paramount, it was not the only salient issue. As Ronald Kline and I argue elsewhere (Lieberman and Kline 2017), electricity also inspired a distinctive and enduring utopianism because it reinforced the hope that scientific methods could unlock usable truths that had been previously unknown. In *Moving the Mountain*, Gilman incorporates this fantasy into a conversation between the narrator and his old friend, Frank Borderson. Frank and the narrator reunite by an electric fireplace, which inspires Frank to explain the recent social changes in a new way:

> Again we sat silent. I ... sat looking into the fire; the soft shimmering play of rosy light and warmth with which electricity now gave jewels to our rooms.
>
> [Frank] followed my eyes.
>
> "That clean, safe, beautiful power was always here, John—but we had not learned of it. The power of wind and water and steam were here—before we

learned to use them. All this splendid power of human life was here—only we did not know it." (Gilman 1911b, 138)

Though *Moving the Mountain* does include new and improved electric devices, it is an electric hearth—a slightly modified version of an ancient tool—that sparks this moment of meditation.[40] Frank's comment "That clean, safe, beautiful power was always here … but we had not learned of it" offers insight into why electricity held such fascination for Gilman and her contemporaries. In this moment, electricity once again serves as a metonymy for other energy sources. The multiplying applications of electrical power signal the possibility of finding and tapping other heretofore-unknown energy sources, such as human batteries or (as in the passage just quoted) the "splendid power of human life."[41]

The other salient aspect of electrical utopianism is, of course, utopianism. The social meanings of electricity are necessary but not sufficient interpretive contexts for understanding the fictional and nonfictional imaginaries that I chronicled in this chapter. However powerful these writers supposed electricity or other inventions to be, they all also agreed that narrative could provoke the changes they hoped to see. After all, they employed pens and typewriters rather than (or in addition to) soldering irons in their pursuit of social change.[42] Technical experts and utopians shared the conviction that they could write the plural potentialities of electrification into a singular outcome. Industry insiders mounted a public-relations campaign to advocate private ownership of electrical utilities. Politicians like Pinchot wrote policies that promoted public ownership in response. Gilman wrote treatises to expand the purview of women's work, and she aspired to replace fantasies of individualism with those of interdependence. Even Steinmetz, the electrical engineer, invested some hope in the power of narrative. In *America and the New Epoch* (1916) he called for a new literature of interconnectedness, suggesting that innovative narratives would be necessary to ignite social change. He argued that the new epoch required "the assistance of those numerous writers who are not connected with corporations nor with the muckraking crowd" because "there is within the huge modern industrial corporation a wonderful field of romance and interest, still unknown and untouched by any writer, which in the hands of a Kipling or a Jack London would give most wonderful stories, more interesting and fascinating than any of the tales of bygone ages of the world's history."

Steinmetz and Gilman agreed that a new owner's manual for electricity should be written. And they appealed to others to join their cause by telling new stories about the radical interdependence that electricity might promote. Although individualism was not so easily displaced in the American cultural imagination, my next chapter will show that Jack London did alter his "call of the wild" style to tell a different story about the yet-unrealized possibilities of electrification—just as Gilman and Steinmetz had hoped.

4 THE CALL OF THE WIRES: JACK LONDON AND THE INTERPRETIVE FLEXIBILITY OF ELECTRICAL POWER

Although he is best remembered today for his semi-autobiographical man-versus-nature fiction, Jack London also wrote extensively about electrification.[1] What could this adventurer possibly have to say about these technical developments? To understand his investment in this lesser-known project, we must consider London before he was known either as a writer or as a socialist. This history begins years before he would write *The Call of the Wild*. It begins, instead, when London pursued a much different calling—when he tried and failed to become an electrician.

In a memoir and treatise against alcoholism titled *John Barleycorn*, London recalls the moment when he first decided that he wanted to be an electrician. He had been working in jute mills, and he found the job as degrading as it was unprofitable. After nursing his indignation for weeks, he resolved to follow another pursuit: "The jute mills failed of its agreement to increase my pay to a dollar and a quarter a day, and I, a free-born American boy whose direct ancestors had fought in all the wars from the old pre-Revolutionary Indian wars down, exercised my sovereign right of free contract by quitting the job. I was still resolved to settle down, and I looked about me. One thing was clear. Unskilled labour didn't pay. I must learn a trade, and I decided on electricity" (London 1913a, 187). This nonchalant decision seems whimsical, even arbitrary. But as he continues his reminiscence, London reveals why he saw electricity as a beacon that could guide him out from poverty. Like many Americans in the Gilded Age, he hoped that the field of electrical development would fulfill the unmet promises of the machine age.

In *John Barleycorn*, London conveys the disparate connotations of machinery and electricity by employing different narrative conventions to describe them. He characterizes his time working in the cannery and jute mills as automated, dehumanizing. Specifically, he describes his "beastial life at the machine" with repetitive diction that emphasizes the numbing sameness of his every day. In contrast, he frames his misadventure in the electric power plant quixotically. London invites his reader to hope, as he once did, that this field offered new opportunities to industrious workers. He outlines his romantic expectations with the ironizing distance of hindsight: "Any boy, who took employment with any firm, could, by thrift, energy, and sobriety, learn the business and rise from position to position until he was taken in as a junior partner. After that the senior partnership was only a matter of time. Very often—so ran the myth—the boy, by reason of his steadiness and application, married his employer's daughter." Certain that he could become an electrician by working his way up from the bottom, London "bade farewell forever to the adventure-path, and went out to the power-plant of one of our Oakland street railways" (187–188).

A precocious young London arrives at this power plant unannounced to solicit a job. There he meets with "the superintendent himself, in a private office so fine that it almost stunned me" (188). The Oaklander initially interprets this "stunning" display of wealth as evidence that he, too, could climb from rags to riches in this facility: clearly money could be made in this profession. Excited by this vision of prosperity, London applies for a job with the earnest enthusiasm that would later be his hallmark: "I told [the superintendent] I wanted to become a practical electrician, that I was unafraid of work, that I was used to hard work, and that all he had to do was look at me to see I was fit and strong. I told him the I wanted to begin right at the bottom and work up, that I wanted to devote my life to this one occupation and this one employment" (188).

From this point on, London's description of his interview with the superintendent toggles between a seemingly unembellished account of their dialogue and parenthetical asides that underscore his own naiveté. Two of London's asides are "as I listened with a swelling heart, I wondered if it was his daughter I was to marry" and "By this time I was sure that it was his daughter, and I was wondering how much stock he might own in the company" (189). These interjections situate London's electrical aspirations in terms of "the myths which were the heritage of the American boy" (187). He expects the happy, sentimental conclusion that was common in dime novels of the day. With hard work—a burden he was already accustomed to—wealth, status, and domestic bliss could be his!

Much to London's happy surprise, the superintendent does offer him a job. And it is "right at the bottom," just as London requested. This much of the "myth" is accessible. But the fantasy of working his way up from this position proves untenable. After six months of shoveling coal to fuel the power plant, with no hope of advancement or technical training, London learns that the superintendent's wealth represents the strategic exploitation of the "American boy" mythology rather than an actualization of that dream. He explains this epiphany with two striking sentences: "I thought he was making an electrician of me. In truth in fact, he was saving fifty dollars a month operating expenses to the company" (193).

London's account reads as a conventional fall from innocence to experience. It registers the failure of romantic and machine-age conventions for describing labor in the age of rapid electrification. London thought he could pull himself up from poverty in this emergent profession. Yet by the 1890s, when he sought the job, the field of electrical expertise had become specialized and corporatized. The skills that were once learned through apprenticeship or on-the-job training were now more often credentialed in colleges or technical schools.[2] As the historian of engineering Edwin T. Layton Jr. argues (1971, 4), "By 1870, there were twenty-one engineering colleges, but only 866 degrees had been conferred. College education became increasingly common" for engineers of all types, including electrical engineers.[3] The field had changed, but the story had not. And according to London that fact was useful to capital.

This was a life-changing realization for London, who repeated this anecdote in several forms throughout his life. Still, despite the shock of his disappointment, London always treated this experience exceptionally. Though his toil in the power plant would be as menial and physically exhausting as his work in the cannery and in jute mills, he consistently described the former experience in uniquely wholesome terms. In a letter to his future wife, Charmian Kittredge, London remembered his time in the power plant with some fondness as a time when he "felt clean—clean inside" (London 1988, 392). This assertion of cleanliness is astonishing considering that London spent his powerhouse days shoveling coal into a sooty furnace. By describing himself as "clean inside," London glamorizes strenuous physical labor. He also calls up the narratives I discussed in chapter 3, which situated electricity as invisible, healthful, and utopic.

London's stint in the power plant taught him a lesson he never forgot about capitalism, but it didn't change the way he felt about electric power or the labor needed to generate it. Like the utopians I mentioned in chapter 3, London believed that electric power was simply in the wrong hands. And he wrote several once-popular but since-forgotten texts that sought to rectify that fact, including the best-seller *Burning Daylight* (1910) and his personal favorite, *The Valley of the Moon* (1913). Despite the relative dearth of critical writing on these novels, both were popular at the time of their publication. In fact, London's contemporaries interpreted *Burning Daylight* and *The Valley of the Moon* as welcome turns away from the author's overtly socialist literature.[4]

Burning Daylight and *The Valley of the Moon* depict the electrification of the American West from the perspectives of the system builders and the working class, respectively. In the former novel, the hero Elam Harnish (nicknamed Burning Daylight) builds power plants in the Yukon and across the Sierra Nevada region. In the latter, the lower-class Anglo-Saxon protagonists (Billy and the transparently named Saxon) abandon Oakland to work on their own electrically irrigated farm plots. These novels were published three years apart—a significant timespan for an author as prolific as London. Nonetheless, their similar trajectories convey London's investment in the alternative form of power generation that he idealizes in both novels. In this chapter, I focus primarily on *Burning Daylight* because it offers interesting insight into the reification of electrical systems. But both *Burning Daylight* and *The Valley of the Moon* explore different possibilities for American electrification. After illustrating how electrical systems can exacerbate class tensions in the context of the early-twentieth-century city, both novels conclude with pastoral scenes of domestic power generation in Sonoma Valley—the idyllic place that London calls "the valley of the moon."

I resurrect these novels and read them in conversation with the industry's electrification narratives for two reasons. First, they illustrate what historians of technology call "interpretive flexibility" of electrical systems.[5] That is, they suggest that Americans might enjoy electrical power without having to buy it from central stations. The Social Construction of Technology scholars who coined and use the phrase "interpretive flexibility" expanded upon the humanist's use of the verb "interpret": to SCOT scholars, *interpretation* connotes the full, embodied understanding of an artifact—the way people use or design it—in addition to the meanings literary scholars glean from texts.[6] Although London didn't use this vocabulary, he did draw attention to choices Americans might make to enjoy electrical power without kowtowing to corporations or monopolies. In so doing, he demonstrated that the trajectory of electrification was not preordained. Its future was still flexible. London frames these alternate choices in socio-political and economic terms. He poses a question that resonates with one that Donald MacKenzie (1996, 43) asks in a different context: "Do capitalists (or men) merely abuse machinery for their own purposes, or do those purposes somehow shape the machine?"

Second, these novels lend to our understanding of London's legacy. Literary scholars have overlooked the electrical dimension to London's work because they have tended to downplay his engagement with themes we call *technological* today. That means that Jack London's critics (like Charlotte Perkins Gilman's critics) have not employed the technological fallacy as often as Mark Twain scholars have. It also means that they have overlooked some of the nuance of his depictions of electricity and other such systems. Although little has been written about *Burning Daylight* or *The Valley of the Moon*, the scholars who have taken up these texts have criticized each novel's sentimental conclusions. For example, Jonathan Auerbach and Christopher Hugh Gair have argued, reasonably, that both purport to solve real-world problems with impossibly fantastic conclusions.[7] Those critics are right: *fin de siècle* Americans could not pick up and relocate to affordable, usable valleys in this historical moment. After all, both novels take place after the much-discussed "closing of the American frontier."[8] But by reading these novels in the context of the history of electricity, other interpretations of electrification and of London's novels become available. Specifically, I argue that his alternative models of power generation were attainable, even if his pastoral settings were not. London's conclusions remind readers that "there is not just one possible way or one best way of designing an artifact," as the SCOT scholars

Trevor Pinch and Wiebe Bijker have argued.[9] In both novels, London suggests that readers could choose to produce their own power instead of buying it from corrupt companies like the one that exploited him. In that sense, London was right: the development of high-power, long-distance transmission systems was not inevitable.

These novels also expand our understanding of London's body of work and of the genre he is most closely associated with today: literary naturalism. In today's literary histories, this genre is often defined as the subset of literary realism that explains human life scientifically in terms of "evolution" and "force." London certainly engages with these themes; like the other authors we now label naturalistic, he cites Herbert Spencer and relishes in descriptions of animalistic strength. Yet, as Nancy Glazener argues, analysts' use of the category "literary naturalism" can be misleading, because writers like London didn't use that phrase to describe themselves.[10]

To scholars seeking out coherent works of literary naturalism, *Burning Daylight* and *The Valley of the Moon* appear flawed and incoherent. Their pastoral conclusions seem like deviations from the naturalistic mode rather than well-crafted narrative choices. And that isn't the only reason these novels have remained out of print. To the general reading public, these novels pose a related challenge. When we spend our days writing and reading in fluorescent-lit offices, we want writers like London to transport us into the terrifying, thrilling, frosty outdoors and not into the design room of a power plant. No wonder *The Call of the Wild*, *White Fang*, and "To Build a Fire" have received asymmetrical attention from readers for generations.

Although London's canonical works may answer our own "call of the wild," we could gain insight and perspective by re-calibrating our assumptions about this prolific and popular writer. If we place London within a broader intellectual and cultural history, we can examine another way that the histories of literature and technology inform one another. Scholars of science and technology studies have sought evidence of "interpretive flexibility" by reopening controversies among technologists and scientists; I demonstrate that close reading fiction is another viable method for locating different interpretations of systems and artifacts. Understanding London within a broader intellectual and materialist history can also complicate our received literary taxonomies. If we read

Burning Daylight and *The Valley of the Moon* on these terms, then we can better appreciate the hybridity and dynamism of London's rhetorical choices.[11]

THE MYTH OF THE AMERICAN BOY

To appreciate the poignancy of London's electrification narratives, we must first understand the "myth" that duped him in *John Barleycorn*—the "myth" that he eventually strove to write against. This "mythology" was fostered by a popular strand of boys' literature, beginning with Horatio Alger's *Ragged Dick* (1868), a concise coming-of-age story about a young bootblack rising into the middle class with a mix of luck and Puritanical economy. During London's childhood, scores of narratives published for young adults recycled this plot line. These stories collectively cultivated the hope that anyone could change their class status with "thrift, energy, and sobriety." London calls up these formulaic tales when he dreams of marrying his superintendent's daughter after demonstrating his worth on the power plant's floor.

This subgenre evolved in tandem with industry. As small private businesses gave way to professionalized and corporatized markets, new stories began to feature boys in more specialized trades than bootblacking. Dime novels about young inventive geniuses such as Frank Reade Jr. and Tom Edison Jr. demonstrated every week what adventures and rewards "good" boys might expect if they applied their strong work ethic to technical trades. John Trowbridge, the professor of physics and novelist I mentioned in chapter 3, adapted this tradition into a set of hardcover books. In *The Electrical Boy*, Trowbridge drew together the traditional sentimental story line of the boys' novel and technical details about electrical circuitry. In the process, he crafted an adventure novel that was also a guidebook for aspiring electricians. His later novel *The Resolute Mr. Pansy* was a kind of technical *Freedom Writers* of its day; it imagined how a motivated electrical science teacher might reform young ne'er-do-wells into happy, unthreatening members of the middle class.

It seems unlikely that London read Trowbridge's novels specifically, but these fantasies were not limited to the pages of novels, dime or otherwise. There were many places where a young London might have encountered the dream of gaining class status along with electrical expertise. Indeed, the electrical industry exploited this popular fictional mode in its promotional materials. Advertisements put out by the electrical industry celebrated the rise of exceptional individuals such as Thomas Edison or Nikola Tesla, perpetuating the myth that electrical fields offered opportunities to any enterprising man, regardless of his formal education. As I noted in chapter 3, this gendered language is intentional: this type of technical expertise was gendered as masculine in the late nineteenth century. To the young Jack London, the masculinization of this profession might have contributed to its appeal.

Edison particularly seemed to embody the fantasy of rising to prominence from poverty. Comic strips and newspaper articles celebrated "The Wizard's" inventions as if they had been individual accomplishments. As Charles Bazerman argues (1999, 48), "Both writers who communicated in the purple prose of late Victorian journalism and Gilded Age financiers who communicated in green dollars saw the laboratory as a place of private genius, the place where Edison worked his personal magic." In actuality, Bazerman notes, the laboratory "was a place of collaborative work." Still, the construction of Edison as an individual genius inspired writers to turn him into the hero of boy's novels. The public image of the electrical industry and the nascent field of young adult adventure literature coevolved. In fact, works in that subgenre of literature later came to be known as "Edisonades" for their engagement with Edison's public image and all that it came to signify.

These myths about Edison only tell part of the story. Private electrical companies—independently or working together under an umbrella organization, the National Electric Light Association (NELA)[12]—hired press agencies to distribute editorials to periodicals across the country. These editorials were not marked as sponsored material, and they amplified the messages that were circulating simultaneously in the form of advertisements and official press releases. Since this promotional campaign was designed to seem polyvocal, it discussed electrification from multiple perspectives. Their romantic accounts of Edison and others were only one part of a much larger story.

The density of utilities propaganda that circulated in American periodicals—especially after 1920—cannot be overstated. The Federal Trade Commission didn't tabulate numbers for every year that the electrical industry planted these articles, but the evidence it gathered in 1925 was staggering. According to Carl D. Thompson's summary of the FTC's investigation of the electrical industry (1932, 321), "In 1925 the total reproduction of the Hofer Service [just one of the many press agencies employed by utilities companies] was reported as 1,954,398 [column] inches; 17,589,292 lines [of print]; 16,286 newspaper pages." Hofer "prepared articles, editorials, and news items in a service which they supplied to between 13,000 and 14,000 weekly and daily newspapers in all parts of the country" (323).

Many of the articles that were covertly sponsored by electrical interests were overtly political. They supported legislation in which the electric industry had a vested interest, such as the 1920 California Water and Power Act. They also advocated private rather than public ownership of electrical utilities.[13] In the process, they characterized privately owned long-distance transmission systems as economical, efficient, clean, and as the product of American ingenuity. Public officials who distrusted private industry depicted expansive electrical systems in more menacing terms. For example, as he brought a lawsuit against NELA and General Electric, Senator George W. Norris of Nebraska argued that "a gigantic trust that has fastened its fangs upon the people of the United States from the Atlantic to the Pacific and from the Great Lakes to the Gulf" (quoted in Thompson 1932, xvii).[14] He claimed that the privatization of electrical power was not as rational and unproblematic a decision as industry press agents suggested.

According to the record the electrical industry left strewn across the nation's newspapers, the fundamental question concerning electric development was that of adopting private versus public power. But over the course of that extensive campaign, industry promoters contributed to other, subtler conversations that have received little critical attention. For example, in the process of bolstering its larger political agenda, NELA and its predecessors also developed a fantastic image of the western engineer. They attempted to balance nostalgia for bygone times with excitement for nationwide modernization.

One publication that came to be affiliated with NELA, *The Electrical Age*, described the transition from Old West to New as magical and seamless: "We have lost that picturesque vision which rose like a mirage from the high hills and wide-stretching plains of the West; it has passed, with its prairie schooners, bands of whooping indians and devil-may-care cowboys, into tradition and ancient history. The West, in accepting civilization, has undergone enchantment and transformation. It is no longer wild and woolly, but a great centre of industries, healthy development and potential wealth." This July 10, 1897, article, titled "Developments in the West," laments the loss of the native and the "picturesque vision" of the frontier. But it describes western industrialization affirmatively. It likens the "healthy development" of the West to "enchantment." This account of modernization describes electrical development as coextensive with the mythology of the "Old West."

The article goes on to celebrate how the West might offer a new electrical frontier that might re-invigorate and strengthen the nation:

> Those that live in some of our old fashioned eastern cities, on attending this exposition, will realize in surprise how great a spirit of advancement is animating these distant centres; how powerful the incentive that within half a century built up these great monuments to western civilization and, lastly, what force of character, hardihood, vigor, and intelligence were required to lay the foundations of a city in such inimical surroundings. We should congratulate the West for their efforts in strengthening this mighty brotherhood that reaches from ocean to ocean.

Without explicitly mentioning the frontier, this piece evokes stereotypes about the spirit of the pioneer by emphasizing the "hardihood" and "vigor" of the electrical workers who build monuments "in such inimical surroundings." Like many articles about western electrification, "Developments in the West" endorses the popular idea "that the West will redeem the East, instead of the more traditional idea that the East will reform the West in its own image" (Folsom 1989, 91). Specifically, this article echoes Frederick Jackson Turner's "frontier thesis": the notion that the struggle against obdurate western terrain helped to transform European immigrants into "real" Americans. According to Turner (1921, 26), the frontier was wholly American because it defied "the European system of scientific administration." Turner extolled the virtue of the frontier only after it was deemed "closed" by the 1890 census. Although the census claimed that the entire continent had been mapped and settled, the electrical industry preserved the mythology of the West by suggesting that it was home to a new kind of pioneer—the "pioneering" engineer or electrician.

Promotional texts about western electrification drew heavily on the dime-novel mythology of the pioneer. Still, press agents did more than adapt existing myths to the field of engineering: they wrote a history of electrical progress as it was unfolding. They claimed that western conditions contributed to broader changes in the field of electrical engineering. Between 1895 and 1915, western land-grant universities were educating a new class of engineers who—according to industry publicists—faced a different set of problems than their eastern counterparts. An article in the June 1, 1912 issue of *Electrical World* titled "The Transmission Systems of the Great West" exemplifies this theme: "The engineers who have worked chiefly under Coast conditions … have no respect for constituted authority when reliance upon it leads to results less directly than seems desirable. … If an impossible dam has to be erected to store the water for a great transmission, they build it and it stays in place and does its work. … If three or four stations must be operated together in defiance of all precedents, in go the switches, and the plants operate as if they had worked together from the very beginning." As was noted above, the turn of the twentieth century was a time of specialization for electrical engineering. Articles such as the one quoted here push against the trend toward classroom learning by reasserting the importance and the thrill of hands-on tinkering. "Transmission Systems of the Great West" describes western engineers as mavericks who defy the constraining conventions of their chosen field. According to this promotional account, these "pioneering" technicians will not be ruled by "laws"—not by the by-laws of their professional organizations, and not by the theoretical laws of electro-dynamics.[15]

Such articles may have given young Jack London hope that his brute strength and lack of experience could be desirable traits in a western electrician.[16] More generally, they characterized high-power, long-distance transmission systems as the handiwork of a new breed of frontiersman. According to this narrative, western electrical engineers had no choice but to develop long-distance transmission systems because western cities were already established long distances away from moving water and coal mines. This account assuaged regional fears that eastern companies would homogenize the West. It also naturalized the course of electrical development by making the expansion of power networks seem ideal or inevitable.

London was not the only young boy to have been taken in by this misleading blend of rags-to-riches literature and electrical industry promotions. In *Twenty Years at Hull House*, Jane Addams writes: "[W]e have come across many cases in which boys have vainly tried to secure such opportunities for themselves. During the trial of a boy of ten recently arrested for truancy, it developed that he had spent many hours watching the electrical construction in a downtown building, and many others in the public library 'reading about electricity.'" (1912, 441) Ostensibly the "myth" that London discusses in *John Barleycorn* was convincing in its day. We cannot trace the success or failure of Addams's anonymous boy, but London's body of work reveals much about the disparity between these romantic stories and the actual condition of the electrical industry in the early 1890s. He tells a different story about electrification—one that undermines the industry's naturalization and mythologization of the long-distance transmission system.[17]

BURNING DAYLIGHT

London's first novel about electrification, *Burning Daylight*, was originally serialized in the *New York Herald* from June through August of 1910, then published as a novel by Macmillan later the same year. The novel traces the migration of a frontiersman Burning Daylight from wild Alaska to industrialized California. In many respects, Daylight (who is alternately called by this nickname and by his given name Elam Harnish in the novel) is London's typical protagonist. He is physically superior to his indigenous Alaskan travel companions, who are invariably amazed by his energy and speed. After traveling with Daylight for two days, a "worshipful" indigenous Alaskan character, Kama, thinks "No wonder the race of white men conquered … when it bred men like this man" (London 1910, 44).

In contrast to the caricatured Kama, Daylight embodies an ideal "civilized primitivism" that permeates London's Northland stories: he can survive the wild in the "Indian fashion," but he also retains some association with civilization through his whiteness. London defines Daylight's character by emphasizing what he is not: he is not a native, nor is he a "man of soft civilization" who would have "grown lean and woe-begone" on the food that Daylight processes into abundant energy (London 1910,

39). He is strong, Anglo-Saxon, and fantastically attuned to the natural world.

Daylight is also electrifying. He can "send a flood of warmth" through any room he enters—can invigorate the dreary winter darkness so that "nothing languished when [he] was around" (3). Throughout the first several chapters of the novel, he is a power source that illuminates and warms every room he enters. He also has "nerves [that] carried messages more quickly" than other men's, muscles that work with "lightning" speed, and a "a supreme organic excellence" (27). Endowed with superior nerves and reflexes, he embodies "remarkable vitality" and astounding efficiency. When he eats, there is "no waste"; everything is "transformed into energy" (39). These metaphors characterize Daylight as a human powerhouse. Everything he consumes becomes usable energy.[18]

The correlation between Daylight and electricity becomes more pronounced when he builds an electrical mining system near his Alaskan home. He undertakes this electrical project because—and not in spite of—his environmental awareness. Especially attuned to change, he rightly foresees that a gold rush will soon begin in his frontier town. When prospectors descend on the area, he realizes he might do a better job in the "big game" of gold mining because he appreciates and understands the land. Daylight "looked at the naked hills and realized the enormous wastage of wood that had taken place" (117–118). To mitigate the ecological devastation in the area, he constructs an electrical system of mining that preserves trees and minimizes waste. Daylight's decision to build this apparatus draws on the connotations of electricity that London appealed to in *John Barleycorn*. In both his memoir and *Burning Daylight*, London treats electricity as a natural energy rather than as a product of American industrialization or corporatization. In this way, London appeals to fantasies about the West as a naturally electrified space.

Unlike the rags-to-riches mythology that I discussed above, this fantasy about electricity as a natural aspect of the West was corroborated by sources with substantial ethical appeal. Even the United States Geological Survey labeled a perceived abundance of atmospheric electricity as a defining characteristic of the western territories. In the otherwise dry *Sixth Annual Report of the United States Geological Survey of the Territories* (1873), Henry Gannett explains his inspiration to name a mountain in Yellowstone "Electric Peak" in a richly detailed anecdote, which has continued to be reprinted in publicity about Yellowstone for more than a hundred years:

> A thunder-shower was approaching as we neared the summit of the mountain. I was above the others of the party, and, when about fifty feet below the summit, the electric current began to pass through my body. At first I felt nothing, but heard a crackling noise, similar to a rapid discharge of sparks from a friction machine. Immediately after, I began to feel a tingling or prickling sensation in my head and the ends of my fingers, which, as well as the noise, increased rapidly, until, when I reached the top, the noise, which had not changed its character, was deafening, and my hair stood completely on end, while the tingling, pricking sensation was absolutely painful. Taking off my hat partially relieved it. I started down again, and met the others twenty-five or thirty feet below the summit. They were affected similarly, but in a less degree. One of them attempted to go to the top, but had proceeded but a few feet when he received quite a severe shock, which felled him as if he had stumbled. We then returned down the mountain about three hundred feet, and to this point we still heard and felt the electricity. (Gannett 1873, 807)

The electrical atmosphere permeates the surveyors' bodies, challenging their role as objective observers by forcing them to acknowledge their embodiment. Though overwhelming, the electrical currents are neither fatal nor horrifying. Reminiscent of Thomas Jefferson's description of the natural bridge in *Notes on the State of Virginia* (1781), Gannett figures electricity as the material manifestation of America's natural sublime. Later, experts in different fields, including the electro-medical doctor George Miller Beard and the popular journalist George Wharton James, corroborated Gannett's impression of the West as inherently electrical.[19] In this context, Daylight's use of electricity reads as an intelligent use of nature itself, rather than an intrusion of a machine into the untamed Yukon.

At first the narrator emphasizes the naturalness of electricity by minimizing the appearance of the power plant in the text itself. He notes simply that Daylight "built his dredges, imported his machinery, and made the gold of Ophir immediately accessible" (120). The implied ease of extracting gold from the land and the brevity of this passage emphasize the plant's minimal impact on the environment. But a few lines later the narrator subtly raises questions about this invention when he recounts the protagonist's "evolution" from frontiersman to system builder: "he, who five years before had crossed over the divide from Indian River and threaded the silent wilderness, his dogs packing Indian fashion, himself living Indian fashion on straight moose meat, now heard the hoarse whistles calling his hundreds of laborers to work, and watched them toil under the white glare of the arc-lamps" (120). Earlier in the novel, London emphasized the strenuousness of his protagonist's life on the frontier. In fact, Daylight almost dies on several occasions before he undertakes this final Alaskan venture. However, in this passage the repetition of the word *Indian* creates a sense of serenity that contrasts with the cacophonous imagery that concludes this periodic sentence. In Daylight's electrical system, the sounds are "hoarse," the lights "glare," and the workers "toil." Even if the "Indian" life Daylight led was challenging, this passage asks whether the life of a system manager would be preferable. In this account, the modernization of the frontier seems less "enchanting" than it does in descriptions by affiliates of the electrical industry.

In addition to disrupting Daylight's "Indian" rhythms, the power plant threatens to change his relationship to the people of Ophir. The man who was once a friend and neighbor to all now creates a hierarchical system in which "hundreds of laborers" become "his." This rhetorical move calls up a trope that the literary scholar Mary Lawlor describes as common to literary naturalism: it demonstrates that "the classical American narrative of free individuals in free space was a faulty paradigm [because] the energy of empire had made precisely this paradigm impossible" (2000, 45). Daylight becomes a local hero and a mainstay of the national periodical press because he can conquer the wilderness with his invention. In conquering, however, he compromises the very wilderness that predicated his heroic, masculine identity.

Daylight's transformation from frontiersman to industry mogul doesn't follow immediately from his financial windfall. This character, like his author, learns the compromises of capitalism the hard way. As the narrator explains: "The grim Yukon life had failed to make Daylight hard. It required civilization to produce this result. ... The change marked his face itself. ... Less often appeared the playful curl of his lips, the smile in the wrinkling corners of his eyes. The eyes themselves, black and flashing, like an Indian's, betrayed glints of cruelty and brutal consciousness of power. His tremendous vitality remained, and radiated from all his being, but it was vitality under the new aspect of the man-trampling man-conqueror" (London 1910, 162–163). At this turning point in the novel, London layers in the common conventions we now associate with literary naturalism and socialist writing: he emphasizes how capitalism forcefully reshapes the body and mind of an otherwise good-natured man. Made increasingly nervous and cruel by the stress of his daily life, Daylight begins drinking to excess, and he develops a paunch. Throughout these transformational scenes, London emphasizes how the socioeconomic forces of the city incentivize antisocial, misanthropic behavior. Now an individualist rather than a community member, Daylight's "tremendous vitality" still radiates, but with cruelty instead of warmth.

THE FALL AND RISE OF BURNING DAYLIGHT

This change in Daylight's character accords with the writing that literary scholars now identify as naturalistic. Specifically, his gradual decline into a "man-trampling man-conqueror" follows from London's readings of Herbert Spencer. Spencer was a philosophical celebrity of his day; his works offered his wide readership a vocabulary for postulating the relationships among biology, sociology, and (obliquely) physics. Not only did he coin the popular expression *social Darwinism*; he also brought metaphors from biology to bear on sociology, and vice versa.[20] Spencer inspired Jack London, Theodore Dreiser, and other literary naturalists to write about the correlation between atavism and poverty. To these activist-writers, poverty became a "force" that caused humans to devolve. They invoked Spencer as they sought to articulate the diffuse, institutional effects of privilege and oppression.

Much has been written about Spencer's influence on the literary naturalists and particularly on London. But Spencer's influence reached beyond this coterie of writers. Edwin T. Layton Jr. notes that *fin de siècle* engineers developed a distinctive professional ideology that drew heavily from Spencer. Inventors who were not professional engineers were also influenced by Spencer. Thomas Edison purportedly celebrated Spencer's philosophy, as did Nikola Tesla.[21] Thus, when London weaves Spencerian notions of degeneration into his plot, he aligns himself with a broader intelligentsia than the canonical literary naturalists.

I mention the technicians' understanding of Spencer because their interpretation may have exerted more influence on "literary naturalists" than Layton or literary scholars have previously supposed. Literary critics generally cite Spencer as the inspiration for scenes in which characters give way to base instincts (as a crueler Daylight does). Indeed, *Burning Daylight* initially appears to invoke Spencer in this conventional way: it emphasizes the hazards of this character's immersion in capitalism. Readers who were acquainted with Spencer or London would expect Daylight to continue on this downward spiral. This was a trajectory that had been traced before in Frank Norris's *McTeague* (1899), in Dreiser's *Sister Carrie* (1900), and even in London's own *South of the Slot* (1909). Yet London calls up these conventions only to defy them. The protagonist of *Burning Daylight* redeems his humanity by undertaking two more electrical endeavors in the second half of the novel.

According to Layton, *fin de siècle* engineers insisted that invention was a creative process that wasn't shaped by the deterministic forces of Spencer's cosmology. "In discussing social evolution," Layton argues (1971, 55), "engineers emphasized the importance of technological innovation. But to engineers inventions were free creations of the human spirit, not the determinate consequences of natural law." This type of thinking inflected London's body of work. Daylight, like many Spencerian characters, is physically and psychologically manipulated by "forces" such as capitalism. But, like the engineers of his day, he describes his electrical endeavors as exceptions to this rule of "natural law."

Daylight doesn't come to this realization himself. Midway into the novel, he meets a new character who thwarts his descent into misanthropy. Dede Mason, Daylight's stenographer, provokes the hero to think about what he might like to make besides money. His answer is electricity. Unlike London's earlier man-versus-nature heroes, he doesn't long for his wilder days. He misses the fleeting moment when the call of the wild and the call of the wires resonated with one another. He explains that this was the endeavor that brought him the most "creative joy" (London 1910, 254):

> [T]here was Ophir—the most God-forsaken moose-pasture of a creek you ever laid eyes on. I made that into the big Ophir. Why, I ran the water in there from the Rinkabilly, eighty miles away. They said I couldn't, but I did it, and I did it by myself. The dam and the flume cost me four million. But you should

have seen that Ophir—power plants, electric lights, and hundreds of men on the pay-roll, working night and day. I guess I do get an inkling of what you mean by making a thing. I made Ophir and … I sure am proud of her now, just as the last time I laid eyes on her. (255)

As Daylight now misremembers it, the Ophir power plant augmented his natural physical strength and abilities. It allowed him to imagine that he single-handedly created a town from a "moose-pasture." By speaking in the first person ("*I* ran the water there … and *I* did it by myself"), Daylight describes himself as a western folk hero who can reshape the landscape. As in the dime novels of the day, his recollection exaggerates his individual accomplishments. Earlier in the novel, the narrator remarks that Daylight didn't build the entire system alone, but here the narrator omits the engineers who built the machinery that he imported, as well as the workers who "toiled" in it to extract his wealth from the earth.

In the context of London's biography, this remembrance can read like a moment of wish fulfillment. Perhaps the author imagines through the avatar of Daylight what it would be like to control electric power before any larger corporate interests owned the land or the equipment or the laborers. We have seen a similar dream before in Mark Twain's *A Connecticut Yankee*. Alternatively, we might read this passage satirically. Perhaps London pokes fun at the deceptively individualistic narratives that once duped him. In either case, Daylight demonstrates how easily captains of industry could erase technical labor. When he writes myriad workers out of his memory, he separates his electrical endeavor from the collaborative labor that produced and actuated it. He slowly collapses a complex system into one reified artifact that he alone controls.

Once Daylight reframes his Ophir power plant in these hyperbolic and individualistic terms, he decides to construct another power system in his pursuit of "creative joy." Instead of building a local single-purpose system, like his electrical mining plant, he begins to build wide-reaching "electric roads." According to the industry-sponsored mythology, such projects layered a new zone for exploration, inhabitation, and development atop the one-time western frontier. In contrast, London's narrator describes this project matter-of-factly: "At the same time that his electric roads were building out through the hills, the hay-fields were being surveyed and broken up into city squares. … Cement sidewalks were also laid, so that all the purchaser had to do was to select his lot and architect and start building. The quick service of Daylight's new electric roads into Oakland made this big district immediately accessible" (London 1910, 265). The vacillation between passive and active voice in this passage raises questions about who actually executes the labor and construction involved in this large utilities project. Who surveyed and broke up the hay fields? The narrator's shift into the passive voice again obscures the cooperative dimension of electrical development. And this is not the only detail that he glosses. Which legislators approved these projects? How did potential users respond?

This omission of users might have seemed conspicuous to contemporary readers, who had recently witnessed public debates about electric public transportation in urban centers. As Eric Schatzberg argues (2001, 67), Americans who were opposed to overhead wires began raising public objections to this common trolley system in the 1890s. London's use of the term *electric roads* indicates that the wires for Daylight's system might have been built underground. But wire panic was not the only salient concern for turn-of-the-century activists who protested electrical transportation systems. Some "condemned the epidemic of injuries to people (both pedestrians and passengers) who fell under the wheels of streetcars" (Schatzberg 2001, 71); others attempted to "block electrification" because they so despised figures like Daylight, who became wealthy from these enterprises "while the city obtained nothing in return" (71); others simply didn't want to lose their "streets as spaces for social interaction" (84). Although actual "electric roads" were developed amid heated debates or (later) "civic celebrations" (82–83), Daylight omits any external perspectives—as if assenting or dissenting people played no part in the development of new wealth and new material cultures.

As we revisit these descriptions of Daylight's electrical endeavors today, it is important to recognize that the social meanings of his systems were unstable. As a self-proclaimed socialist, London likely intended to criticize Daylight's self-aggrandized and over-simplified account. After all, London had once worked as a laborer who kept power running to Bay Area electric roads. He would have had a vested interest in resisting the invisibilization of such labor. Still, London's description of bloodless—nearly autonomous—electrical development might have corroborated some readers' deterministic understanding of progress. The longer *Burning Daylight* remained in print, the more likely it was that these depictions of electrical development could seem accurate rather than conspicuously dehumanized. In other words, narratives like that in *Burning Daylight* might have unintentionally encouraged the erasure of human labor from discussions of large-scale electrical systems.

Whether we read *Burning Daylight* as criticizing or contributing to the reification of electrical systems, the novel flies in the face of "Developments in the West" and other promotional articles that described western electrification as rational, altruistic, and manly. Parodying these mythologized heroes of the industry, Daylight builds electrical systems selfishly, to augment his wealth and his sense of personal power. He chooses to build a long-distance system not because the public demands it, and not in order to solve a technical or environmental problem, but because he feels personally discontented:

> Not content with manufacturing the electricity for his street railways in the old-fashioned way, in power-houses, Daylight organized the Sierra and Salvador Power Company. This immediately assumed large proportions. Crossing the San Joaquin Valley on the way from the mountains, and plunging through the Contra Costa hills, there were many towns, and even a robust city, that could be supplied with power, also with light; and it became a street-and-house lighting project as well. As soon as the purchase of power sites in the Sierras was rushed through, the survey parties were out and building operations begun. (London 1910, 298).

Detached from the local communities and environments he electrifies, Daylight visualizes this endeavor from a bird's-eye perspective. He never wonders whether his "street-and-house lighting project" will help or harm people in the San Joaquin Valley. While he believes that this project represents creative rather than destructive joy, he conspicuously avoids any discussion of human actors. The passages quoted above fail to mention the people who inhabit and build the systems, or the people who are left out of them. Even the phrase "made this big district immediately accessible" raises the question "accessible to whom?" This omission registers the distance between the producers and consumers under a regime of corporate electrification.

Burning Daylight doesn't focus on this "consumption junction" (Cowan 2012); the novel never examines the consumers who might use the systems the protagonist produces. Instead, it emphasizes how he is harmed by his own preoccupations with power and money. The narrator describes

Daylight's aspirations as dangerously, addictively profitable: "The profit on this land was enormous.... But this money that flowed in upon him was immediately poured back into his other investments" (London 1910, 266). This accretion and re-investment of wealth becomes a compulsion, as Jennifer Fleissner (2004) might describe it—a repeated cycle of capitalist frenzy. Pivotally, Dede recognizes and interrupts this cycle. She refuses to marry Daylight while he remains a captain of industry, protesting that "a wife would be only a brief diversion" to someone with his responsibilities (London 1910, 294). She forces him to choose between the creative joy of building electrical systems and the procreative joy of genteel domesticity. Faced with this ultimatum, Daylight relinquishes his millions and opts to marry.

Before the couple retires to "the valley of the moon," Dede asks Daylight to consider the people who might be affected by his move. She protests: "I know something of the fight you have been making.... If you stop now, all the work you have done, everything, will be destroyed. You have no right to do it" (325). Here, Dede draws some attention to the laborers and consumers who he elides throughout this episode. But Daylight dismisses her concern. He assures her in the abstract idiom of industry: "Nothing will be destroyed, Dede, nothing. You don't understand this business game. It's done on paper" (326). In this passage, Daylight sounds like an early-twentieth-century counterpart of Bruno Latour and Steve Woolgar, arguing that paper is the real motive force that makes his power plant run.[22] Daylight's emphasis on paper also highlights the nexus between writing and electrification, subtly hinting that written words (perhaps including the novel *Burning Daylight* itself) could inflect the course of electrical development. Above all, he reframes the Sierra and Salvador Power Company as an impersonal and rhizomatic system, although he claimed to control and create that system solitarily just a few chapters earlier.

Daylight and Dede's conversation ultimately naturalizes the power system. It suggests that electrification has a momentum of its own. This rhetorical maneuver resonates with (and arguably paves the way for) the rhetorical slippage that Leo Marx identified in today's usage of that "hazardous concept" *technology*. Daylight can describe a large collaborative project as his own *or* as an autonomous system, as if these attributions were trivially interchangeable. Both of the strategies he uses to describe this power system erase the many human decisions and actions that determine its uses and design. This rhetorical move illustrates the logic by which complex systems became reified, even before the word *technology* gained prominence in the American vernacular.[23]

Suggestively, London associates this rhetorical displacement with the scale of the system he describes. In the conclusion to the novel, Daylight builds another power system. In this final sequence, the uses of the generator appear more pronounced than the energy itself. Daylight no longer builds power systems for their own sake; he now builds them with his domestic needs in minds:

> Daylight devoted himself to the lightening of Dede's labors, and it was her brother who incited him to utilize the splendid water-power of the ranch that was running to waste. … Together [Daylight and Dede's brother] installed a Pelton wheel. Besides sawing wood and turning his lathe and grindstone, Daylight connected the power with the churn; but his great triumph was when he put his arm around Dede's waist and led her out to inspect a washing machine, run by the Pelton wheel, which really worked and really washed clothes. (London 1910, 351).

By introducing the brother, who helps to generate power with Daylight in order to "lighten" Dede's load, London re-inscribes emergent narratives about men as producers and women as consumers that I discussed in chapter 3. At the same time, this passage suggests that local generators, also historically called "isolated plants," might strengthen family and community bonds in a closed, controllable setting—even if other forms of power generation alienate producers and consumers in urban settings. London romanticizes this small power system, hinting that personally generated electrical power is less dangerous and more fulfilling than the web of wires that brings this energy from increasingly distant locations. This scene, contrasted with the novel's climax, underscores London's engagement with what we would now call *co-construction*: Daylight builds his own power systems, but they also build him in turn.

THE MANY LIVES OF AN ELECTRICAL INDUSTRY STORY

Revisiting the electrical industry's promotional campaign can help to clarify the intervention London makes with the saccharine conclusion mentioned above. Industry advertisers naturalize the development of long-distance transmission systems; London implies that Americans need not rely on the large-scale system to enjoy washing machines and other conveniences. To comprehend how the industry advocated long-distance power transmission, I pause here to trace the recirculation of one promotional story across three publications. This comparative interlude also demonstrates how technical reports about electrification transformed into mythic narratives about ingenuity and American identity.

Let us begin with a version that enjoyed the largest readership. On February 24, 1896, the *Chicago Daily Tribune* included an article, patriotically titled "Americans Have Great Courage," that applauded the ingenuity of the western engineers who built some of the first long-distance transmission systems. The *Tribune* credits this article to its author, John McGhie, and to the venue in which it originally appeared, *Cassier's Magazine*, an illustrated monthly periodical about engineering. Although no version of this article lists McGhie's credentials or professional affiliation, he was the Manager of the Advertising Department in General Electric's Cleveland division.[24] And General Electric belonged to the nascent umbrella organization, NELA.

In the *Chicago Daily Tribune*, McGhie's article is poorly edited down to four paragraphs that use electrical developments to demonstrate American exceptionalism. This foreshortened version of McGhie's article was also re-published verbatim in *The Electrical Journal* two months later.[25] Interestingly, the original *Cassier's* article was a more complicated sixteen-page study of "Long-Distance Transmission of Power by Electricity in the United States." The re-titling and curtailing of this article demonstrate how easily a technical account of power transmission could be transformed into a puff piece about American courage.

The *Tribune* version offers a whiggish history of development in the field of electrical power engineering. It implies that the transition from direct current to alternating current was logical and inevitable: "[T]he direct current was, perforce, discarded, and the alternating current called into requisition." This claim appears in many contemporary narratives about electrical progress. According to this simplified and linear account, Thomas Edison's DC system was displaced by George Westinghouse's AC system because the latter could transmit electricity across longer distances more efficiently. This narrative implies that the move to high-power, long-distance transmission systems was logical and desirable. This is the very assumption that Jack London toys with in *Burning Daylight* when he raises questions about the protagonist's motivation for developing his extensive power system. Daylight doesn't build his large-scale power system because he is rational; he builds it because he is addicted unhealthily to the accretion of wealth.

The *Tribune* article goes on to characterize the development of large-scale electrical systems as the result of engineers' unique problem-solving abilities: "Attainment of an economical solution was by no means easy. Difficulty after difficulty arose, requiring countless experiments to elucidate; and alteration after alteration in machine was made, involving the expenditure of vast sums. By successive and painful stages a solution was finally reached." This report was misleading. Later historians of technology show that alternating current and direct current systems "existed in a complementary way until the newer system became the dominant one" (Hughes 1983, 15). In fact, John McGhie himself noted this fact in the longer *Cassier's* article from which the *Tribune* piece was drawn. The full article included images of "the direct current side" of "the new power station at Portland, Oregon" (McGhie 1896, 361–362). In abbreviating this report for a broader lay audience, press agents omitted these details. In so doing, they classified long-distance transmission systems as progress from earlier short-distance systems, as if it were always inevitably desirable to transmit power across great distances.

The *Tribune* also omits McGhie's first expository page. In the *Cassier's* version, the writer suggests that developments in long-distance power transmission were equally motivated by diligent technological curiosity and by profit seeking: "For the theoretical electrician ... long distance power transmission offered a fascinating invitation; to the manufacturers of electrical apparatus, it looked an arable land which, with but slight tilling, might laugh in a bountiful harvest of acceptable profits" (McGhie 1896, 359).[26] The *Tribune* article downplays the profit motive. It includes, instead, McGhie's brief rehearsal of utopian notions of endlessly renewable clean water power: "Hitherto unutilized water powers have become, in sanguine imagination, possible gold mines in *futuro*, and the elimination of the domestic coal heap and relegation of the steam engine to the oblivion which awaits the discarded, have become articles of faith with water power proprietors" (360). London mimics this language in his electrification novels, but to a different end. The General Electric–affiliated article suggests that long-distance transmission systems allow Americans to access figurative "gold mines" and to eradicate the "coal heap." In contrast, London suggests that users could derive the same benefits from individually owned and operated generators.

As in the utopian literature I discussed in chapter 3, this article codes these advancements as specifically American. McGhie adds:

> By far the greatest number of the long-distance transmission installations of the world are situated in the United States. The American seems endowed with the courage of temerity, and is willing to adopt a new thing with promise only, where other nationalities demand assurance or proof. A possibility has a special attraction for the American mind, and the risk of its realization is willingly run. It is this spirit that has covered the United States with electric lighting stations, spread a network of electric car lines over every city of any importance in its boundaries, and initiated the supersession of the steam locomotive itself from its main line railways. (McGhie 1896, 360, reprinted in the *Chicago Daily Tribune* and in *The Electrical Journal*)

The narrative voice in these paragraphs retains some critical distance from the notion that long-distance power embodied progress. In the passage quoted above, McGhie interjects the clause "in sanguine imagination" and uses the linking verb "seems" to qualify his own exaggerations. He also makes the American "spirit" the subject of his final sentence, hinting that American national identity preceded and necessitated the spread of electrical power. This rhetorical move allows him to naturalize the move to long-distance electrical power distribution without debating whether these systems should be owned by private corporations or by the public.

The *Tribune* and *The Electrical Journal* end the narrative here. This conclusion depicts the spread of electric lighting stations and car lines across the country as rational and patriotic. In *Cassier's*, however, McGhie goes on to complicate his own nationalistic rhetoric. The next paragraph begins with this sentence: "The first long-distance power transmission plant perhaps on the American continent, was that installed by the old Thomson-Houston Company in 1887 in Guatemala" (McGhie 1896, 360). Shifting attention from "the American mind" to "the American continent," McGhie situates electrical power in a spatial and geopolitical market, though he doesn't indicate whether peoples outside of the United

States have an equal mastery over electric power technologies. The power-transmission plant is the subject of the sentence; Guatemala is simply an environment for the uncomplicated installation of this system by an American-owned company. McGhie doesn't mention whether local labor and expertise were used in the construction of this facility. Still, he goes on to suggest that experiments in long-distance power transmission occurred simultaneously in laboratories in Italy, Switzerland, Germany, France, and the United States before engineers in Germany demonstrated feasibility in 1891.

Despite these gestures toward a transnational account of electrical development, the visual rhetoric of this article characterizes power transmission as a distinctly American invention. In its original *Cassier's* form, the article appears beneath a photograph of power lines across an unidentifiable expanse with the subtitle "Cross-Country Power." Unlike the other photographs in the article that are painstakingly labeled and situated in the text as evidence of developments in specific power plants across the country, "Cross-Country Power" captures a sense of generic national identity because of its lack of specificity. The photograph of power lines alongside a wooden fence could represent almost any part of the rural United States. The other photographs interspersed throughout the article show power plants in Portland, Oregon, in Lowell, Massachusetts, in Silverton, Colorado, and in Folsom, California, hinting that high-power transmission is broadly American and often western.

As we saw above, the idealization of the American West offered a trove of emotionally appealing tropes that advertisers used to glorify the development of these long-distance systems. For this reason, the shortened version of McGhie's article emphasized western electrical development, although long-distance transmission systems were being developed in other contexts. London's *Burning Daylight* demonstrates the limitations of the electrical industry's idealized pioneer/engineer. This novel also reminds readers of a technical choice that the industry conspicuously omits from its circulated and recirculated reports: it suggests that Americans might enjoy electricity that they generate for their own purposes. London's intervention in this conversation is subtle. It would have been easy to dismiss his oddly idealistic conclusion as being poorly or quickly written, had he not demonstrated a sustained investment in this issue by revisiting the theme of electrification in *The Valley of the Moon*.

THE VALLEY OF THE MOON

The Valley of the Moon closely imitates *Burning Daylight*. Like its predecessor, it amalgamates naturalistic and pastoral generic conventions. In view of their similarities, I focus here on the issues that the former novel raises but the latter does not. The two novels differ primarily in their depictions of class. Whereas Burning Daylight is a captain of industry, *The Valley of the Moon*'s protagonists are both of the working class. Establishing this class-based theme, London opens this latter novel in a laundry in Oakland, where Saxon, the heroine, is ironing shirts under electric lights (London 1913b, 7). As she moves from her electrically lit workspace into her dimly gaslit home, the narrator implies that "electrical progress" has done nothing to help the working class. Saxon cannot afford the shirts she presses or the electric power she uses to iron them. Her inability to access the wealth of the West is the central problem that this novel explores; the narrator presumes her entitlement to the land.

Saxon parallels Dede Mason. She is a working-class, pragmatic, and waifish white woman who is irrepressibly drawn to the physical strength of the novel's hero, Billy. Billy, in turn, resembles Daylight. He has the physical prowess of a pioneer, though he has spent his whole life in the city. The novel begins with their courtship and marriage. Then, like many of the novels we now call naturalistic, it describes how the "forces" of industrial-age America tear their relationship apart. After a series of misfortunes, including a violent labor dispute, Billy's arrest, and Saxon's miscarriage, Saxon chooses to remove herself from Oakland's economy entirely. She forages for mussels and driftwood in the bay, and she carries her found food and firewood home. London exaggerates her eschewal of modern socioeconomic systems by describing how Saxon strategically avoids electric light: "She sought the darker side of the street at the corner and hurried across the zone of electric light to avoid detection by the neighbors" (271). At the border of this "zone of electric light," London contrasts Saxon's self-reliance with the image of her longtime friend, Mary.

Like Saxon, Mary's life has been shattered by a recent clash between her husband's union and his employer. However, Mary chooses to live under the lights that Saxon resolutely avoids. This ancillary character becomes a sex worker to remain in the capitalist economy. The setting of this scene alongside the pivotal labor dispute suggests a correlation between sex work and other forms of physical labor: Mary's decision to sell sex aligns with laborers' decision to sell their bodies for wages. Insofar as one of the corporations with union-busting policies is Niles Electric, Mary's scene (and the first half of the novel more generally) depicts electric light as a part of a larger problem. Electricity in this urban setting is a metonym for the dehumanizing logic of capitalism. It transforms the street into a stage where neighbors can observe each other's suffering without helping one another.

Like Dede in *Burning Daylight*, Saxon refuses to participate in the destructive, compulsive behaviors of urban life. She will not engage in an economy that devalues her work and harms her body. After Billy is released from prison, Saxon insists that they "chuck Oakland" (277). Following in the tradition of their pioneering parents, Saxon and Billy set out from their home by foot to find free government land. Their trek across the state re-invigorates their bodies and their relationship. It also reveals how much California has changed since their parents colonized the land a generation earlier. As the writer and historian Gerald Haslam argues (1989, 136), California was engineered from a desolate into a fertile landscape between the 1860s and the 1890s, with correlated social and agricultural results: "Damming those rivers and channeling their waters for irrigation helped convert the valley into the world's richest agricultural region. In turn, that richness has attracted an ethnically and socially diverse series of migrants to work its fields: Chinese, Japanese, Italian, Portuguese, Sikh, German, Filipino, Mexican, black, and various poor whites, among others." The agricultural practices of these largely segregated populations dismay and amaze the protagonists along their journey.

In the country, Billy and Saxon meet a series of farmers who help acquaint them with the modern conditions of California agriculture. They are especially impressed with Benson, a recent college graduate with a degree in agricultural engineering. He argues that soil is more valuable than gold, pointing out that the "gold's gone, and the ... soil remains." Throughout this episode, Benson repeatedly alludes to contemporaneous debates about the stability of the gold standard.[27] He essentially advocates a return to a Lockean model in which value is produced by mixing labor with the land. This land-based fantasy particularly appeals to Saxon, who had been looking for an alternative socioeconomic model to wage-labor capitalism.

The farmer-mentor's argument is two pronged. First, he claims that the current economy of the United States is predicated on an unsustainable and insatiable cycle of chasing wealth. He summarizes the current condition of American cities in this complaint: "Those of us who haven't anything rot in the cities. Those of us who have land, sell it and go to the cities. Some become larger capitalists; some go into the professions; the rest spend the money and start rotting when it's gone, and if it lasts their life-time their children do the rotting for them" (368). Benson's speech retroactively justifies Billy and Saxon's decision to leave the city. The image of "rot" recalls the naturalistic plot line of decay.

Second, Benson characterizes the pastoral dream of living off the land as feasible and not nostalgic. To accomplish this, he uses his knowledge of modern agricultural techniques to explain why a return to the land might work in 1913, even though previous generations of white settlers failed to survive as farmers. According to Benson, Anglo-Saxon claim jumpers (including Billy's and Saxon's parents) overvalued gold and lacked the expert knowledge that was needed to run a bountiful farm. Benson states his case this way: "Why, in France, I've seen hill peasants mining their stream-beds for soil as our fathers mined the streams of California for gold. Only our gold's gone, and the peasants' soil remains ... growing something all the time." This explanation astounds Billy: "'My God!' Billy muttered in awe-stricken tones. 'Our folks never done that. No wonder they lost out. ... [But it] was our folks that made this country,' Billy reflected. 'Fought for it, opened it up, did everything—.'" Benson finishes Billy's sentence: they did everything "but develop it." Benson continues: "We did our best to destroy it, as we destroyed the soil of New England" (366–367). This conversation enables Billy to rationalize why his ancestors lost their land, despite their purported racial superiority. At the same time, the farmer's snipe about New England characterizes the Northeast as infertile and irreversibly industrialized. It implies that the West still offers the promise of an electric pastoral that already seems impossible elsewhere in the United States.

This exchange also suggests that the history of violent colonization is the only connection that binds Saxon and Billy to land that they consider their rightful inheritance. Indeed, Billy's complaint that his ancestors "fought for it" doesn't seem particularly convincing. The agricultural engineer suggests that modern farming techniques—specifically electrical irrigation systems—could help these nativists develop a better-founded claim to this land. In that sense, Benson demonstrates the promise of using *technology* in the old sense of the word—to mean knowledge of specialized techniques. Still, it is worth noting that London doesn't use that word in this novel; Thorstein Veblen's book *The Instinct of Workmanship*, which would go a long way toward reifying *technology*, would not come out for another two years. London, like the writers I discussed in previous chapters, continues to focus on electricity specifically rather than on technical or mechanical progress in general.

To distinguish between modern American farming techniques espoused by Benson and the strategies employed by non-white farming communities, London constructs a symbolic system that codes immigrants' labor as difficult and ugly, while describing electrical irrigation as nearly magical. For example, Billy describes the "Porchugeeze" as "livin' like a pig" (303). In contrast, the narrator portrays an electrified farm plot as Elysian:

> Stepping into a small shed, [a modern farmer] turned an electric switch, and a motor the size of a fruit box hummed into action. A five-inch stream of sparkling water splashed into the shallow main ditch of his irrigation system and flowed away across the orchard through many laterals.
>
> "Isn't it beautiful, eh?—beautiful! beautiful!" [the farmer] chanted in an ecstasy. " … It makes a gold mine laughable." (459)

The repetition of "beautiful" emphasizes the allure of this marriage of electrical power and irrigation. Although London associated electricity with the exploitative capitalist economy earlier in the novel, the orchard's small-scale grid represents an ideally beautiful and healthful modernity. Again, the comparison with gold pokes fun at capitalistic value systems. Read alongside *Burning Daylight*, this passage suggests that agricultural applications are more desirable than mining. Although electricity could be used to extract wealth from the land in both cases, gold is not renewable once it is depleted. In contrast, London underscores that the beauty and bounty of the fields can be replenished. Electrical irrigation allows Billy and Saxon to transform from wage laborers into happy, modern homesteaders. This plot twist aligns London's lengthier novel with the utopias I examined in chapter 3.

The Valley of the Moon revises the electric pastoral that London developed in *Burning Daylight*. Both novels promote the fantasy that the complex city system might be broken into closed, controllable circuits that form the foundation for healthier white families or communities. Published at a moment when American cities were coping with widespread neurasthenia and poverty, they both suggest that interactive and thoughtful electrical applications might solve various social and technical issues. But the central issues of these novels diverge. *Burning Daylight* focuses on the dehumanizing logic of capitalism. Capitalism remains an issue in London's later novel, but the central problem of *The Valley of the Moon* is race.

London's earlier novel depicted isolated plants as raceless, labor-saving artifacts. Although Dede and Daylight are both white, the conclusion to his 1910 novel leaves open the possibility that anyone might re-interpret electrical power use in the same manner. *The Valley of the Moon* depicts electric irrigation as a white power in both senses of the word: as a power source controlled by Anglo-Saxon experts and as one way these experts could reassert control over non-white others and the land. Above, I argued that technical accounts of long-distance power transmission could all too easily be converted into a patriotic narrative of American courage. The same could be said of the transition from *Burning Daylight* to *The Valley of the Moon*: the ease with which London modifies a story about an electric pastoral into a strikingly similar story about white power demonstrates how closely aligned these discourses were in the American cultural imagination at the time.

Race becomes an issue in *The Valley of the Moon* because corporate-owned power stations create crises of racial and national identity for London's characters. Billy and Saxon cannot understand how some recent immigrants can thrive in modern California while the well-meaning white poor toil uselessly in the cities. To articulate this perceived injustice, London describes Billy, Saxon, and other descendants of California's white settlers as "the last of the Mohegans," whose land is unfairly infringed upon by comparatively recent immigrants and capitalists (1913b, 155). This analogy to the "dying race" of Native Americans represents a departure from *Burning Daylight*. London compares the hero of his former novel to an Indian, but only to describe this character's outdoorsy way of life and not to describe his racial identity.

London's use of the "Mohegan" moniker in *The Valley of the Moon* problematically suggests that Billy and Saxon have a right to the land because their ancestors violently colonized it first.[28] This nativist language conjures *The Electrical Age*'s mythic account of the Old and the New West.[29] According to the electrical industry's press agents and to other Anglo-Saxon writers of the time, "whooping Indians" had been displaced from the Old West by the introduction of modern inventions. London's appropriation of the "Mohegan" name and his depictions of electric power draw their emotional appeal from this common myth: it hints that the same fate might await white urban workers if they don't gain control of this new and "powerful" invention.

ALTERNATIVE FUTURES

As I mentioned above, few critics have examined *The Valley of the Moon* or *Burning Daylight*. Those who have analyzed them have emphasized the improbability of this domestic withdrawal into an electrically enhanced wilderness twenty years after the official closing of the American frontier. For example, Christopher Hugh Gair argues:

> By moving to a nearly self-sufficient, Jeffersonian rural idyll, the protagonists Elam and Dede Harnish gain a degree of moral agency impossible in the determined worlds of the city—where much of the novel is set—and of naturalist fiction. This transformation is problematic for various reasons, the most important of which are: first, that the Harnishes' escape from the market requires money earned in that market ... and, second, the move depends upon a retreat into history and into the economic structure of the nation at least fifty years earlier. (1996, 141)

Gair perceptively observes that both novels vilify capitalism but still illogically rely upon it. Indeed, Lewis Mumford noticed this dissonance decades before Gair did, and even went so far as to claim that London was no socialist at all, stating that "he brandished the epithet socialist as a description of himself and his ideas" while being "gullible enough to swallow Kipling's doctrine of the White Man's Burden, believed in the supremacy of the Nordics, who were then quaintly called Anglo-Saxons, and clung to socialism, it would seem, chiefly to give an additional luster of braggadocio and romanticism to his career" (Mumford 1968, 126). *The Valley of the Moon* especially exemplifies these inconsistencies: its conclusion requires Billy and Saxon to sell their electrically irrigated crops to a broader market. In that sense, the protagonists' relocation doesn't solve the market logic of capitalism that London (at least superficially) denigrates throughout the novel.

Still, I disagree with Gair on his second point: neither *Valley* nor *Burning Daylight* advocates a "retreat into history," despite London's affinity for Jefferson's yeoman republic. Literary historians' overdetermination of London as a call-of-the-wild author underwrites this reading. Today's readers expect London to want to return to nature. Thus, when his characters move to Sonoma Valley, they seem to realize the pre-industrial fantasy that underwrote London's earlier fiction. However, both novels imagine an alternative future instead of a return to the past. Both were written as long-distance power systems were being developed by engineers and by a near-monopoly of electrical concerns. By turning (not re-turning) to an alternative form of electrical power use, both novels demonstrate that Americans had more choices than the two that the electrical industry highlighted: in addition to asking whether electric power should be publicly or privately owned, potential consumers still might have asked whether electricity generated locally for personal use might better serve their purposes than central-station power.

London emphasizes the practicability of this choice in *The Valley of the Moon* by contrasting the problems he perceives in electrical power generation to related problems he finds in electrical communication. Early in Saxon and Billy's sojourn, they encounter an "American" telephone lineman who complains "here am I, workin' for the telephone company an' puttin' in a telephone for [a man] from the Azores that can't speak American yet" (1913b, 311). To the lineman and to the narrator, modern communication systems connect and enfranchise the wrong Americans. For London, new immigrants and the over-civilized rich benefitted unfairly from exploited white, working-class labor. This lineman's complaint resonates with the novel's earlier depictions of labor power struggles and corrupt power companies. Neither the telephone system nor Oakland's power system has any regard for Saxon and Billy's nativist claims to the land. Crucially, London doesn't propose an alternative communication system in the novel; electric power generation is the only social and technical problem he purports to solve.

London treats the issues of electric communications and electric power divergently because these systems had distinctive interpretations, in both the colloquial and the SCOT sense of the word. By the time London wrote *The Valley of the Moon*, communication and transportation systems were relatively "closed." By juxtaposing the unsolvable problem of the telephone to the (supposedly) solvable problem of electric power generation, *The Valley of the Moon* implies that electrical systems were still flexible even if electrical communications systems were not.[30] As London hinted, consumers still could choose to use isolated plants rather than long-distance power. In the years before the Rural Electrification Administration lobbied for a national grid, many electrical consumers in rural areas made precisely that decision.

Recognizing the interpretive flexibility of electrical systems, as London does, is an important step toward avoiding the technological fallacy. If we acknowledge that specific artifacts and systems such as electrical power generators could have evolved in multiple ways—if we question the received notion that long-distance power transmission was the most rational response to generic technical problems—we challenge the nebulous autonomy of that concept *technology*. We take away its implied agency.

INDETERMINATE PRESENTS

London's idealization of isolated generators didn't stall the development of large-scale systems. In fact, readers might have enjoyed his fiction precisely because it allowed them to escape briefly from the realities of corporate industrialization that alarmed the writer himself. But, as I suggested in my introduction, such texts can be valuable for the questions they help us to ask, rather than the ones they seem to answer. In this context, the rhetorical strategies London uses are as interesting as the alternative forms of electrification he promotes. *Burning Daylight* and

The Valley of the Moon emphasize the human aspect of human–machine interactions. To the electrical industry, the West was an emotionally appealing image to exploit. To London, the West was a lived-in space with resilient local histories and prejudices that could not be erased by electrical development. Concomitantly, London's novels demanded that technical neologistic verbs take a preposition: the West was not just electrified—it was electrified by and for individuals with wide-ranging (and often competing) economic and interpersonal interests, which inflected how (and why, and whether) these systems would be integrated into lives and landscapes.

In addition to emphasizing the human decisions that shape electrical development, London employed a strategy I have previously highlighted in this study: pastiche. Although his fictions project hope for America's electrified future, they reach resolutions to intricate, realistic problems only by cobbling together romantic, naturalistic, socialistic, and capitalistic imagery. At the level of content, London's protagonists never confront their reliance on the exploitative systems they claim to escape. At the level of form, both of his novels abruptly shift from a naturalistic trajectory into pastoral resolutions. Such jarring changes in tone and genre hint at the limitations of literary realism and naturalism in attempting to account for the contingent and nonlinear interactions of economies, politics, individuals, and material power systems. As such, these texts make conspicuous the struggle to develop vocabularies and narrative forms capable of describing complex interdependencies and nonlinear, systemic interactions.

The hybridity of London's writing may be an aesthetic flaw, but it is a poignant rhetorical strategy all the same. As Lewis Mumford would insist a generation later:

> The romantic movement was retrospective, walled-in, sentimental: in a word, regressive. It lessened the shock of the new order, but it was, for the greater part, a movement of escape.
>
> But to confess this is not to say that the romantic movement was unimportant or unjustified. On the contrary, one cannot comprehend the typical dilemmas of the new civilization unless one understands the reason and the rationale of the romantic reaction against it, and sees how necessary it is to import the positive elements in the romantic attitude into the new social synthesis. Romanticism as an alternative to the machine is dead: indeed it

never was alive: but the forces and ideas once archaically represented by romanticism are necessary ingredients in the new civilization, and the need today is to translate them into direct social modes of expression. (Mumford 2010, 287)

Knit-together narratives such as *Burning Daylight* and *The Valley of the Moon* strove for just such a synthesis of romanticism and utilitarianism. These novels problematized the electrical industry's claim that electricity meant only one thing: progress. London resists this interpretation. He demonstrates that users can define what "progress" means for themselves. For a time the character Burning Daylight believes that new power systems embodied progress, but ultimately he decides that locally generated power could better enable him to thrive as a person. To Saxon and Billy, the large-scale power system never represents progress; it is an agent of their oppression. But they are not opposed to electrification. They find progressive applications of electricity that also serve their personal desires. These story lines hint that the power system does not in itself represent modernity or progress—the real power lies in how people choose to design or use it.

Although London has long been classified as an anti-modern writer, it is this kind of ambiguity—not anti-modernity—that pervades his body of work. We can see this indeterminacy in most of the narratives he published, including texts that have little to do with electrification. I find the most arresting example in London's quasi-biographical magnum opus, *Martin Eden*, in which he paints a beautiful but indefinable scene in electric light: "Then they fell upon each other, like young bulls, in all the glory of youth, with naked fists, with hatred, with desire to hurt, to maim, to destroy. All the painful, thousand years' gains of man in his upward climb through creation were lost. Only the electric light remained, a milestone on the path of the great human adventure" (1908, 135). This moment, when London turns from his signature descriptions of raw violent life to acknowledge the electric light, challenges me. His tone, heavy with irony, raises a question: What does "progress" mean in the face of atavistic wildness? How can the electric light symbolize an "upward climb through creation" when so many humans still wallow and fight in the mud?

The light in this passage is an image of existential absurdity. It resists the meanings that Americans commonly invest in it. This light is not the legible beacon of uplift that London pursued when he sought a job in electricity. It is not the emblem of a new frontier that industry promoters purported. The light is unintelligible in this moment because its development was always couched in narratives of progress—narratives that appear nonsensical when we are made to remember that humans are never just *homo faber* (tool users). They are also always other things—including irrational tangles of impulses and instincts. The incongruousness of these two narratives of human life would not have been lost on a Spencerian philosophe like London. In addition to identifying actual "interpretive flexibility" in power generation, then, London's novels invite us to wonder how we might interpret the electric light in general. What does this invention really mean to the modern American, if it cannot faithfully signify *progress* or *civility*? And who determines those meanings?

5 RALPH ELLISON'S AND LEWIS MUMFORD'S ELECTRIFYING HUMANISM

The questions that Jack London raised in *Martin Eden* continued to concern writers and public intellectuals in the generation after the author's untimely death: How could electricity represent social progress in the face of human brutishness? How could any invention? In this chapter, I explore how two particularly prominent public figures[1]—Lewis Mumford and Ralph Ellison—reframed these questions. These writers didn't ask whether electricity or other inventions were progressive; they wondered how writing could help to make them so. In order to construct genuinely reformative social meanings of electricity, both became students of history. They believed that America's past held the key to its most promising future.

I choose a comparison of these two visionaries as my final case study for two reasons. First, they explicitly and insightfully responded to the literary and cultural history that I have chronicled thus far. As the historians of technology Thomas P. Hughes and Agatha C. Hughes argued (1990, 3), Mumford's reflections are uniquely valuable because "he sought a usable past" that could help him make sense of his present moment. Ralph Ellison shared that aim. Each writer aspired to learn from literary and cultural history in order to cultivate "the spiritual resources that would sustain the benign values that would in turn reshape technology for creative and constructive, not destructive, ends" (Hughes and Hughes 1990, 9). Each contended that well-crafted prose could inspire Americans to rethink their relationship to electricity and other inventions. And, since both writers used the word *technology* carefully and sparingly, they modeled how to avoid the fallacy that I discussed in previous chapters by retaining a deliberate critical distance from this "hazardous" but now-ubiquitous term.

Second, Ellison criticized Mumford with an incisiveness that demands interdisciplinary consideration. This fact has been under-examined because Mumford and Ellison were sorted into presumably disjointed disciplinary traditions. Mumford was among the first American nonfiction writers to study the interplay of technology and society (Molella 1990, 22). As such he became a foundational figure for historians of technology, and his literary criticism has received comparatively little attention. Meanwhile, Ellison's fiction and nonfiction have been incorporated into literary and cultural history. Even his discussions of technology have seemed to fall under the exclusive purview of literary scholars, enabling cultural historians of technology to overlook his insights. While scholars in science and technology studies and literary studies have pored over Mumford and Ellison's contributions, respectively, Ellison's reading of Mumford has received little attention from scholars in either field.[2]

In this chapter, I demonstrate that these historical figures can be better understood together. I argue that Ellison attends to issues that Mumford oversimplifies: the novelist's masterpiece, *Invisible Man*, illustrates how American race relations underwrite the social meanings and uses of electricity and other technologies. By telling this story, Ellison fosters a more nuanced philosophy of technology and humanism than the one that Mumford proffers in *The Golden Day* (1926) and in *Technics and Civilization* (1934). But that is not to say that Ellison's voice should displace Mumford's within this tradition. Ellison helps readers to locate the enduring appeal of Mumford's most evocative early writings; reciprocally, Mumford offers an illuminating interpretive context for re-examining Ellison's depictions of electricity, technology, and twentieth-century American life. The intersection that I analyze in this chapter may constitute just a fraction of Ellison's and Mumford's larger bodies of work, but it also illustrates the rich interactions that become invisible when we segregate literary and technological histories. By reading Ellison and Mumford together, I hope to illustrate that—at least to a certain extent—the history of technology is also the history of literature and the history of literature is also the history of technology.

MUMFORD, ELLISON, AND THEIR INTERSECTIONS

Lewis Mumford entered the American intellectual scene in the 1920s, about ten years before Ellison's first short stories were published. A self-taught public intellectual, Mumford cultivated a non-academic perspective intentionally; his writerly persona "was grounded in a principled rejection of the prevailing empirical, scientistic ideology of American universities, with its ideal of detached, context-free or 'objective' knowledge" (Marx 1990, 166). Unrestrained by disciplinary conventions, Mumford researched and wrote extensively about literature, sociology, utopian studies, architectural history, and the history of technology. For the purposes of this study, I am particularly interested in Mumford's earlier works that Ralph Ellison read—particularly his pre–World War II studies of literature, technics, and society.[3] The first of these texts was *The Golden Day*. Published only a year after the publication of *An American Tragedy* and ten years after Jack London's death, *The Golden Day* was a paean to American romantics. In it Mumford claims that Ralph Waldo Emerson and Walt Whitman had bolstered a healthy appreciation of machines while continuing to nurture the human spirit. In contrast, he accuses post–Civil War writers of subordinating broader spiritual concerns to realistic representation (as Stephen Crane had done in "The Devil's Acre").

By raising questions about humanism, literature, and technical progress, *The Golden Day* establishes ideas that would later evolve into the more detailed and better-remembered *Technics and Civilization*. In both studies, Mumford argues against the idea that mechanical progress would beget social progress, claiming: "No matter how completely technics relies upon the objective procedures of the sciences, it does not form an independent system, like the universe: it exists as an element in human culture and it promises well or ill as the social groups that exploit it promise well or ill. The machine itself makes no demands and holds out no promises: it is the human spirit that makes demands and keeps promises" (Mumford 1934, 6).[4] In such passages, Mumford's influence on the social history of technology appears palpable. Like the academic historians of technology who would emerge (and honor his contributions) in the following generation, Mumford strives to supplant deterministic narratives about technical development with more nuanced discussions about human and technological interactions.

In the passage quoted in the preceding paragraph and throughout his early work, Mumford carefully characterizes "the machine"—a term he preferred to *technology* for describing the entire system of artifacts, techniques, and arts—as a product of human culture. He needs to clarify this point because of the shift in the notion of progress that I discussed in my first chapter. In the early nineteenth century, machines seemed a clever means to improve American life or democracy. By the early twentieth century, the conflation of technical development and progress had enabled many Americans to imagine that technology (or related plural concepts, including technics and machines) acted as a driving force in human history—as if technical artifacts or systems were extrinsic forces that could change society. Leo Marx and Merritt Roe Smith (1994, xi) characterize this changing conception of progress as a form of "technological determinism" through which "a complex event is made to seem the inescapable yet strikingly plausible result of a technological innovation."[5] Throughout his career, Mumford challenges the simplistic assumption that technical development was a distinctive characteristic of modern human life or a force that stimulated social change. In fact, he often suggests that the perception of change was itself overstated, insofar as neither machines nor social mores were as new as most of his compatriots seemed to believe.[6]

Mumford's critique of this form of determinism sheds light on another facet of the technological fallacy that I addressed in previous chapters. Earlier, I described scholars' tendency to project the concept *technology* onto the past as if it had always existed. Mumford explains why that might be. He claims that Americans see technology everywhere because they have been trained to value mechanical and electrical development over their own emotionality or humanity. According to Mumford, Americans have come to believe that technological progress represented the defining characteristic of human civilization. As a result, they have ceased to appreciate the artistic aspects of their culture and, concomitantly, the multidimensional experiences of human life.

To advance this point, Mumford argues that the prevailing cultural fascination with machines emerges with attendant anti-humanist values: "The direct reaction of the machine was to make people materialistic and rational: its indirect action was often to make them hyper-emotional and irrational. The tendency to ignore the second set of reactions ... has unfortunately been common in many critics of the new industrial order: even Veblen was not free from it" (1934, 28). This cultural critique remains relevant today. We continue to witness a culture war between instrumentalism and humanism, or, as we more commonly frame the dualism today, between STEM fields and the humanities—between the fields that would produce new technical knowledges and applications and those that would help us interpret (among other things) the stakes of and the reasons for such developments. According to Mumford, Americans should not aspire to be less expressive and more scientific in their thinking, as many engineers, politicians, and industrialists claimed (and continue to claim). It would, he suggests, be more worthwhile to develop social and technical systems that would serve Americans' social and emotional needs. Subtly in *The Golden Day* and explicitly in *Technics and Civilization,* Mumford discredits the binarizing logic that glorifies utilitarianism and denigrates romanticism. He recommends that Americans synthesize these sensibilities.[7]

Ralph Ellison shares a resonant understanding of technological and social progress—one that is inspired by reading Mumford closely and critically. Both writers seem to believe that technological development could be promising if and only if it could be calibrated to serve human emotional needs. The novelist summarizes his Mumfordian understanding of technology in a 1958 interview that is published in his 1964 collection *Shadow and Act* as "Some Questions and Some Answers." His interviewer asks a leading question, presuming that Ellison (or any other black man) would oppose technological development: "What is the role of modern industrial evolution on the spiritual crisis of the Negro people?" Ellison responds by undercutting the interviewer's premise and asserting that technology might positively benefit "the Negro people":

> Ironically, black men with the status of slaves contributed much of the brute labor which helped get the industrial revolution under way; in this process they were exploited, their natural resources were ravaged and their institutions and their cultures were devastated, and in most instances they were denied anything like participation in the European cultures which flowered ... under the growth of technology. But now it is precisely technology which promises them release from the brutalizing effects of over three hundred years of racism and European domination. (1964, 264)[8]

Ellison doesn't specify how technology would fulfill this promise or what might constitute "release." He simply acknowledges this outcome as a hopeful possibility. Such responses reveal Ellison's radical attitude toward technology. At a moment when the still-crystallizing concept was increasingly affiliated with white, masculine expertise,[9] he undermines the idea that African-American people were resistant to technology.

Notably, however, Ellison never invests entirely in the ascendant determinism of his cultural moment. He modifies this answer about the transformative potential of technology by adding: "It is not industrial progress per se which damages peoples or cultures, it is the exploitation of peoples in order to keep the machines fed with raw materials. ... There is, I believe, a threat when industrialism is linked to a political doctrine which has as its goal the subjugation of the world" (1964, 265). With this qualification, Ellison challenges the predominant correlation between social and technological progress. He refuses to believe that new inventions will liberate or harm people. Rather, Ellison reminds his interviewer that the products of industrialization serve an ambiguous "role in the lives of people of any racial identity" (264)—a role shaped by the institutions and powers that control them. Dissatisfied with the idea that objects or systems would solve human problems, Ellison and Mumford both maintain that inventions such as electrical power systems could serve a progressive purpose only when developed in tandem with an appreciation of history and a robust humanist ethic.

Elsewhere, I have called this stance toward social and technical progress *technological humanism*.[10] Indeed, both Ellison and Mumford are humanists in the multiple senses of the word. They value humankind's spiritual and emotional growth over material gains or dogma.[11] Both suggest that American life had stagnated—rather than progressed—as a result of a national overemphasis on mechanical development. In fact, both have been labeled anti-modern for taking these positions. But neither imagines that these outcomes are inevitable. Ellison and Mumford are technological humanists because they foster the hope that lay citizens could mindfully reshape the uses and meanings of electrical power and other systems in order to serve humanistic ends. They aspire to supplant the technological determinism of their day with a more nourishing and human-centered understanding of progress.

Although Ellison agrees with Mumford's claim that Americans should resist the inclination to settle for mechanical progress in lieu of meaningful social change, the novelist's perspective is not interchangeable with his contemporary's. Indeed, Ellison once quipped that, when a white intellectual claims "This is American reality," "the Negro tends to answer (not at all concerned that Americans tend generally to fight against any but the most flattering imaginative depictions of their lives), 'Perhaps, but you've left out this, and this, and this. And most of all, what you'd have the world accept as me isn't even human.'" (1964, 25) Although Ellison was discussing realistic fiction in this passage, the same could be said about his response to Mumford's histories and philosophies.

In fact, Mumford's omissions are as conspicuous to twenty-first-century readers as they were to Ellison's hypothetical African-American skeptic. In *The Golden Day*, the nonfiction writer laments the outcomes of the Civil War: "As it turned out, the war was a struggle between two forms of servitude, the slave and the machine. The machine won, and the human spirit was almost as much paralyzed by the victory as it would have been by the defeat. An industrial transformation took place over night: machines were applied to agriculture; they produced new guns and armaments; the factory regime, growing tumultuously in the Eastern cities, steadily undermined the balanced regimen of agriculture and industry which characterized the East before the war" (1926, 67–68). Throughout this passage, Mumford insinuates that slavery and mechanization are equally dehumanizing. This exaggeration lends urgency to his cultural critique, but it also dismisses the human dignity of emancipated African-American slaves.

Ellison takes offense at this. He satirizes Mumford's interpretation by using the "The Golden Day" in his novel *Invisible Man* as the name of a run-down bar in which forgotten veterans, failed professionals, sex workers, and social outcasts gather. In a private letter, Ellison explains his decision to lampoon Mumford's book in this way:

> It wasn't that I didn't admire Mumford. I have owned a copy of the sixth Liveright printing of THE GOLDEN DAY since 1937 and own, and have learned from, most of his books. I was simply upset by his implying that the war which freed my grandparents from slavery was of no real consequence to the broader issues of American society and its culture. What else, other than sheer demonic, masochistic hell-raising, was that bloody war all about if not

slavery and the contentions which flowed there-from? As a self-instructed student, I was quite willing for Mumford to play Aeschylus, Jeremiah, or even God, but not at the price of his converting the most tragic incident in American history into bombastic farce. For in doing so he denied my people the sacrificial role which they had played in the drama. (quoted in Wright 2005, 159)

As Ellison contends in this letter, actual human slavery cannot simply be elided for emphasis. Slavery was not a rhetorical device. It was a real institution with lasting effects that continued to influence the lives of every American.

Mumford's inadequate depiction of servitude and the Civil War in *The Golden Day* cannot be dismissed as a young writer's folly. The metaphor of the machine-enslaved white laborer is an integral aspect of his social and technical vision. Even in his more expansive *Technics and Civilization*, he repeatedly relies on the metaphor of slavery. In *Technics and Civilization*, machines appear as slaves and as masters: "the ordinary workman has the equivalent of 240 slaves to help him," but machines also reduce that same workman to be "dumb, servile, and abject" (Mumford 1934, 324–325). Worse still: the worker is "properly extruded from mechanical production as slave" (414). And yet, his fascination with the slave image notwithstanding, Mumford neglects to explore American race-based slavery—even when he discusses the theft of resources from Africa to fuel the European desire for increased power (281).[12] In this context, Ellison's critique remains relevant as scholars from literary studies and technology studies continue to weigh our intellectual inheritance from both figures.

Still, despite his shortcomings, Mumford inspired many great thinkers who succeeded him, including Ellison. As the fiction writer admits in the letter quoted above, he "owned [and] learned from, most of [Mumford's] books." The literary scholar John S. Wright (2006, 150) summarizes Ellison's revision of Mumford this way: "It was not, then, technology as a vast alienating system of machines moving through history with implacable force that preoccupied Ellison [as it did Mumford]. It was technology as an extension of human lives, as something *someone* makes, *someone* owns, something *some* people oppose, most people *must* use, and *everyone* tries to make sense of." Where Mumford writes about the American cultural imagination in general, Ellison discusses the specific imaginings of dynamic and relatable characters. By focusing on the points where large-scale systems interface with individual lives, Ellison illuminates dimensions of American social and technological life that Mumford's more distant approach inherently overlooks. By studying the complementarity of these perspectives, this chapter sheds light on both writers—and it contends that their narratives remain pertinent to our thinking about technology and society today.

MUMFORD'S ELECTRICAL UTOPIANISM

To appreciate Mumford's nearly utopian affection for electrical power—and Ellison's engagement with those ideas—we must first understand how the social meanings of electricity had evolved in the time between Gilman's book and Mumford's. By the 1920s, in a few far-flung parts of the country—including Mumford's New York home—electricity had already come to seem a natural part of modern life. Nye describes a historical moment when this paradigm shift became apparent: "during World War I, when the United States Fuel Administration turned off the electric signs as an economy measure," New Yorkers petitioned until the lights were re-ignited. They complained that the city was "drab" and "gloomy" without the Great White Way (Nye 1992, 60).[13] Although the majority of the country had yet to be wired for power, electrification had come to seem an expected consequence of "progress."

This sense of inevitability emerged from the fact that electrical systems were stabilizing after a generation of rapid and haphazard change. As Thomas Misa argues (2004, 128), "Technologists in this era [1870–1930], especially those hired in science-based industries, increasingly focused on improving, stabilizing, and entrenching existing systems rather than inventing entirely new ones." While these systems were becoming ensconced in urban and coastal America, the word *technology* was developing new connotations; the term was increasingly used to signify material artifacts that were created with scientific techniques, rather than the study of techniques more generally. Such changes in material culture and in the American vernacular developed in feedback, appreciably changing the social meanings and uses of electricity.

This meaning-making process was also inflected by a new nationwide public-relations campaign. In addition to the NELA promotions that I discussed in chapter 4, General Electric began promoting the idea of "electrical consciousness" shortly after the Great War. This advertising platform was designed to convince Americans that they needed electricity. An internal General Electric document, signed by vice president J. G. Barry, invites employees to imagine a time "when every American begins to ask, 'How can I turn over more of *my* hard work to electricity?'" Barry announces that General Electric will stop selling individual products in order to promote the concept of "electrical progress" instead. "We are not concerned here primarily in selling goods but in promulgating an Idea," he proclaims. "As that Idea takes firmer and firmer hold upon the popular imagination every member of the electrical industry will profit, General Electric with the rest."[14]

This initiative—spearheaded by GE's president, Gerard Swope—consolidated the image of the electrical slave that I mentioned in chapter 1. Promotional materials from this campaign included catch phrases such as "Its Shoulders Never Tire." Aside from the fact that Ruth Schwartz Cowan (1983) demonstrated that these labor-saving devices actually made "more work for mother," rather than relieving work in the manner that they advertised, this promotional campaign might appear innocuous. But by the time Mumford began publishing books and essays about neotechnics, the electrical industry had been drawing analogies to slavery for more than ten years. For example, in 1915 GE ran an advertisement, "Slaves of Yesterday and Today," that juxtaposed images of the slaves who built pyramids with images of electrical machinery, conspicuously eliding the recently emancipated African-American slave. Advertisements in that campaign boasted that electricity minimized the human toll of slavery, making claims such as this: "The Great Pyramid was built, according to Herodotus, by the bitter toil of 100,000 men for twenty years. Men died like flies. The world *does* progress. A modern skyscraper goes up in a fraction of a year—work-men furnishing the skill and General Electric motors the muscle." But the omission of the American slave from such images bolstered the fantasy that the nation's industrial modernity was realized by machines, rather than by exploited, dehumanized, and enslaved people.

The June 1922 issue of the *General Electric Review* extends the logic of this image—a fact which is particularly notable because this document was circulated internally and not used to sell devices to consumers. An

Ellison's and Mumford's Electrifying Humanism 177

Figure 5.1
"Slaves of Yesterday and Today" (General Electric Advertisement, ca. 1915). Reproduced with permission of miSci, Museum of Innovation and Science, Schenectady.

essay titled "Is Democracy to Fail?" asserts the economic benefits of slavery with a vague history lesson. In so doing, it demonstrates that the idea of the electrical slave was not simply a dead metaphor used to describe certain devices:

> Why is it that in the older democracies they could have what we cannot have? It is a strange fact that there is just one point about the old successful democracies that is always forgotten. They had slaves to cultivate their fields, dig their ditches, wash their plates and dishes and to do all those things that no one wants to do. These slaves had no part in the democratic life of the state. They were the captives of their many successful campaigns, and it should be noted that it was not until they had these slaves that they could devote their time to that literature and art which has made them famous throughout the ages.
>
> Yes—it is absolutely true that we are in a bad way in our modern democracies because we have no slaves. (Hewett 1922, 331–332)

Placing slavery under the purview of generic ancient cultures, this essay re-codes the dehumanizing institution as an economic boon and a cultural necessity. Like GE's earlier advertisement, it evades the shameful history and the residual effects of slavery in the United States, conspicuously omitting racism and violence from its historical overview. With only a cursory mention of the human toll of slavery, the essay states that "our modern democracies will not stand for slaves of the old type—the human slave—but they cannot last and flourish without slaves of the new type" (332–333). This language naturalizes slavery to promote electrical devices. By noting that modern democracies will not stand for the indignity of keeping humans as slaves without explaining this change in public opinion, this article rationalizes the history of the practice. In so doing, it insinuates that these "slaves of the new type" enabled the eradication of human slavery, rather than the unrelenting activism of escaped slaves and abolitionists, or the publication of influential narratives, or changes in public policy, or civil war.

The controversial idea of the electrical slave recurred across American letters because it appealed to a distinctively national fantasy. It promised to reconcile the country's slave economy and its corporate economy. It encouraged Americans to look forward rather than reflect backward—to imagine that slavery could be improved just as any outdated device or system could. Such messaging helped electricity appear to be the most rational solution to social and personal problems. In the 1930s, the federal government corroborated this impression that electricity could improve American life by forming the Rural Electrification Administration. By the middle of the twentieth century, elected officials and industrialists were co-contributing to the fantasy that electrical progress spurred social progress.

Although they have disparate responses to the notion of electrical progress or to the use of slavery in General Electric's advertisements, Mumford and Ellison are both drawn to the prospect of widespread electrification. In fact, their bodies of work attest to the enduring appeal of the aesthetic tradition I have archived in this book: both writers remain enthralled by the poetry and possibility of electricity—even as other metonyms for modernity (including electronics, technology, information, and eventually nuclear power) emerged on the American scene.[15]

Mumford is particularly taken with the potential applications of this energy.[16] Whether or not he was familiar with General Electric's campaigns, he echoes some of their promotional language. Like the proponents of the "electrical slave," Mumford's *Technics and Civilization* displaces the recent memories of slavery by attributing it only to "older civilizations," including ancient Greece (279) and Rome (41). He also credits specific inventions with the abolition of slavery: "Thanks to the menial services of wind and water, a large intelligentsia could come into existence, and great works of art and scholarship and science and engineering could be created without recourse to slavery" (118). Passages such as this insinuate that slavery might remain a necessary evil if wind and water could not be harnessed. For Mumford, a culture that does not foster "art and scholarship and science and engineering" would be tantamount to slavery. He goes so far as to claim that "the inhuman exploitation of chattel slavery [is] hardly less inhuman [than the] exploitation of wage slavery" (75).

Slavery, for Mumford, remains an abstraction; he is evidently more comfortable discussing the figurative emancipation of workers than the legal emancipation of American slaves. In fact, throughout *Technics and Civilization* Mumford repeatedly describes the electrical liberation of figuratively enslaved laborers in terms that he drew from the Scottish scientist and writer Patrick Geddes,[17] suggesting that "neotechnic" inventions (including electricity and alloys) and values (including a commitment to lightness and beauty) could liberate modern men from the undesirable aspects of the "paleotechnic" (coal-burning industrial) era.[18]

In Geddes's idiom, Mumford predicts the dawning phase of American social and technical development with ecstatic language: "Light shines on every part of the neotechnic world: it filters through solid objects, it penetrates fog, it glances back from the polished surfaces of mirrors and electrodes. And with light, color comes back and the shape of things, once hidden in fog and smoke, becomes sharp as crystal" (1934, 245). This passage stitches together the discourses of his utopian predecessors with those of his modernist contemporaries. Like the former, he envisions a radiant future in which electricity and other inventions will be used to improve the undesirable aspects of industrialization. Like the latter, he aestheticizes the accouterments of the neotechnic revolution, so that even electrodes come to seem alluring. Yet, even as Mumford describes the dramatic changes that were then being and would later be wrought by neotechnics, he insists that inventions such as electrical power would not recast the modern world in their own image—to Mumford, that would constitute another form of machine enslavement. A few pages later, he assures readers with more conventional aesthetic sensibilities that electricity will improve, rather than displace, traditional art forms by enabling the "constant lighting of the sculpture and the canvas" (336).

While emphasizing light and energy to glorify the coming technical age, Mumford simultaneously admires the material substructure that has made electricity available. He praises the water turbine and the gasoline-powered isolated plant that make it practicable to generate electricity in rural areas. He also celebrates the large-scale power system: "With larger central power stations there are other advantages. Not all power need be absorbed by the local area: by a system of interlinked stations, surplus power may be transmitted over long distances" (1934, 223). Mumford denigrates the technocrats who control large plants. Still, he treats central stations as exceptional for their ability to transmit power across long distances. In this way, he idealizes electricity in all its forms, muddying the distinction that London had drawn between personally and centrally produced electricity.

While celebrating the ingenuity of electrical infrastructure, Mumford also sings the praises of electrical science. He claims that "Clerk-Maxwell's unification of electricity and light is perhaps the most outstanding symbol" of the neotechnic phase (1934, 245). He also extols the new possibilities that electricity has revealed within the sciences: "During the neotechnic phase, the sense of order became much more pervasive and fundamental. The blind whirl of atoms no longer seemed adequate even as a metaphorical description of the universe. During this phase, the hard and fast nature of matter itself underwent a change: it became penetrable to newly discovered electric impulses" (217). He imbues electricity with an aura of agency in these passages. From his perspective, electricity doesn't merely reveal phenomena that were hitherto unseen, as Gilman had claimed. Instead, electricity changes the "nature of matter itself." These passages reveal the extent of Mumford's fascination with electricity—a fascination that threatened to undermine his more critical interpretations of this energy and of "the machine" in general.

These overzealous depictions of electricity can be attributed to what Rosalind Williams (1990b, 47) identifies as Mumford's "lifelong quest to articulate the distinction between 'good' machines and 'bad' ones, and to explain how both the liberating and the repressive ones have emerged in the same history." In *Technics and Civilization*, Mumford explains why neotechnic inventions such as electricity have yet to realize their progressive potential: "Not alone have the older forms of technics served to constrain the development of the neotechnic economy: but the new inventions and devices have been frequently used to maintain, renew, and stabilize the structure of the old order" (1934, 266). This qualification reintroduces his anti-determinist critique. According to Mumford, cultural values would have to change before the neotechnic phase could fully come into being. Consequently, although he uses the past tense when stating that "during the neotechnic phase, the sense of order became much more pervasive and fundamental," Mumford also insists that the neotechnic phase would elude a technologically deterministic society that waited for new inventions to drive social change.

Still, even as Mumford tempers his enthusiasm for electricity in the passage quoted above, he invokes industry publicity. Consider "I Am Electricity," a 1932 advertisement from the "electrical consciousness" campaign. Like Mumford's characterization of the neotechnic phase, the ad depicts electricity as constrained by outdated (paleotechnic) tools and ideas. Specifically, it features an enslaved personification of electricity who pleads "Don't keep me in chains—put me to work!" The accompanying text exploits stereotypes of the shiftless slave by stating "Don't keep me chained in idleness." Although that stereotype was historically racialized as black,[19] this image whitewashes the iconography of American slavery by coding this slave as white-skinned, straight-haired, and eager to serve. Harkening back to the erasure of labor that I discussed in chapter 3, this ad promotes the use of electricity ("tamed" by majority-white engineering firms) to replace work that was often performed by lower-class non-white people.

Technics and Civilization and "electrical slave" advertisements sugarcoat the history (and the very concept) of enslavement in order to promote electrification as a means of attaining personal happiness or social progress. What distinguishes Mumford's account from its industry analogues is his emphasis on humanism. Mumford hopes that his predictions of a better, more beautiful world—juxtaposed with a keen critique of his era—might inspire his audience to do more than adopt electricity into their lives. He seems to want them to reconsider the meaningfulness of their lives altogether: "the positive fostering of the life-conserving occupations and the discouragement of those forms of industry which decrease the expectation of life … all this awaits a culture more deeply concerned with life than even the neotechnic one, in which the calculus

Figure 5.2
"I Am Electricity" (General Electric Advertisement, 1932). Reproduced with permission of miSci, Museum of Innovation and Science, Schenectady.

of energies still takes precedence over the calculus of life" (1934, 248–249). In this passage, Mumford shifts from speaking as an augur of the neotechnic phase to speaking as a technological humanist. He reveals that his ultimate goal is not an electrical age such as General Electric promoted; it is a new social paradigm that values life above all else and designs its technics accordingly. This curt proposition—easy to miss amid dozens of pages of electrical utopianism—is the underlying idea that Ellison would elaborate upon several years later.

TECHNOLOGICAL HUMANISM IN ELLISON'S "BATTLE ROYAL" SCENE

Like Mumford, Ellison was curious about electricity, radio, and other inventions associated with the second industrial revolution. In fact, both writers had formative experiences tinkering with radios and other small electronic devices,[20] and much later Ellison became an early adopter of the computer.[21] But Ellison was more impressed with the aesthetics than the applications of electricity. Building upon the literary tradition that I discussed in previous chapters, he invoked electricity because it could signify on multiple levels simultaneously.[22] Fascinated with ambivalence, Ellison once claimed that "the essence of the word is its ambivalence, and in fiction it is never so effective and revealing as when both potentials are operating simultaneously, as when it mirrors both good and bad, as when it blows both hot and cold in the same breath" (1964, 25). Electricity offered him a vehicle for describing such coexisting potentialities.

Capitalizing on the specific ambivalences of this energy, Ellison repeatedly crafted stunning depictions of electric currents in his fiction. In the process, he defibrillated the dying metaphors of his predecessors. When his Invisible Man steals electricity from Monopolated Light & Power in order to feel "vital aliveness," he reveals a new "body electric" that is distinct from the versions Whitman and Gilman had sung.[23] To his literary mentors, electricity was a universal anatomical inheritance. To Ellison, electricity was never merely natural. It was always distributed through literal and figurative power lines. By artistically exploring that truth, Ellison transformed an increasingly commonplace energy into a striking call for action and reflection. In view of Ellison's emphasis on ambivalence, his technological humanism can be more difficult to discern than Mumford's. In the parlance of his most famous nameless protagonist, Invisible

Man, Ellison broadcasts this idea on "the lower frequencies" (581), amid other messages about society, electricity, and technology. Ellison insinuates where Mumford explicates, rendering their conversation asymmetrical but nonetheless compelling.

While Ellison crafts beautiful depictions of electricity, he appears much more skeptical of this energy than Mumford. In fact, the novelist suggests that Americans' hypertrophied faith in electrical progress serves a similar function to white supremacy. Both ideologies give the white majority a false sense of control over the modern world; both invite Americans of any race to buy new-and-improved gadgets instead of building wise-and-improved selves. And yet, as I indicated above, Ellison could never be classified as a Luddite. An acolyte of Mumford's, Ellison also implies in his fiction and argues in his nonfiction that technical systems founded in humanist thought might actually improve American daily life—even if existing social and technical systems were deeply flawed.

Ellison simultaneously encodes these competing understandings of electricity into the only novel he completed during his lifetime, *Invisible Man*. In fact, one of his most perspicuous depictions of electricity and white supremacy appears in the "battle royal" scene in the novel's first chapter. This scene serves as an especially compelling example because it was published as a short story before Ellison completed *Invisible Man*, and it continues to circulate independently in literature anthologies. When one reads it apart from the larger work, it represents electricity as a hegemonic tool. When one reads it in the embedded contexts of the novel and of Ellison's engagement with Mumford, more ambivalent and compelling interpretations become available.

In this ostensibly anti-modern scene, white patriarchs force young black men to fight one another and to suffer electric shocks for money. The protagonist had been invited by these white men to accept a college scholarship and to deliver a speech, but once he arrives these grotesque white men compel him to partake in the battle. Invisible Man recounts the pain and humiliation of being forced to scrounge for coins on an electrified floor: "I lunged for a yellow coin lying on the blue design of the carpet, touching it and sending a surprised shriek to join those rising around me. I tried frantically to remove my hand but could not let go. A hot, violent force tore through my body, shaking me like a wet rat. The rug was electrified" (27).[24] The protagonist had attended the event to deliver a speech about "the very essence of progress" (17); what he found was shocking proof of social stagnation.

Though Mumford avoids questions of race, he does describe a similar type of corrupted authority in *Technics and Civilization*: "Paleotechnic purposes with neotechnic means: that is the most obvious characteristic of the present order. And that is why a good part of the machines and institutions that boast of being 'new' or 'advanced' or 'progressive' are often so only in the way that a modern battleship is new and advanced: they may in fact be reactionary, and they may stand in the way of the fresh integration of work and art and life that we must seek and create" (1934, 267). The white patriarchs in Ellison's "battle royal" scene do in fact treat electricity like a battleship: they weaponize it to fortify their social position.

If the white men represent the endurance of paleotechnic values, then the coins on the electrified rug signify the failure of the socialist and utopian dreams that I explored in previous chapters. In Ellison's scene, capitalism, racism, and electrical power seem inextricably wired together. As the literary scholar Johnnie Wilcox explains (2007, 1000), "In the Battle Royal, electricity establishes connections between the constituent elements of the ad hoc networks which generate capital, affecting the kinesthetic behavior of connected organisms in such a way that those organisms are identified as racially inferior, as black." Put another way, the "battle royal" scene sets the protagonist's powerlessness against the combined electric and social power of a white establishment. Where earlier writers projected that electricity might improve American democracy by breaking down these hierarchies, Ellison agrees with Mumford that this energy, alone, could never initiate such a change.

Still, even as Ellison criticizes this collusion of electric and white power, his protagonist doesn't hold these power relationships as fixed. Notwithstanding his youth and inexperience, Invisible Man strives twice to subvert his subjugation. First, he "discovered that he could contain the electricity" (27). This "containment" seems more figurative than referential; it isn't clear what "contain electricity" means. Nonetheless, Invisible Man seems to recuperate some composure by reframing the electric current as something he can choose to take into his body, instead of a shock that is forced through him against his will. He also tries a more direct approach: he attempts and fails to pull one white spectator onto the rug.

The first strategy seems to succeed while the second explicitly fails. Allegorically, these outcomes map onto the book's conclusion. The character learns that he cannot "tackle" white power, but that he can subvert it by re-imagining his relationship to electrical and social power.

The impression that the protagonist might exercise some autonomy under these violent conditions is heightened when we consider this scene within its broader novelistic and interpretive contexts. In *Invisible Man*, the "battle royal" scene follows a gripping prologue that previews the novel's conclusion. In the prologue, the narrator confides that he will eventually hide underground and that he will steal the power he needs to play his phonograph and to wire 1,369 electric lights with purposeful inefficiency. He indicates that his entire ceiling will be covered in light by the end of his journey. And he intimates that this act of illegal illumination was only the beginning: "An act of sabotage, you know. I've already begun to wire the wall." To what end? The narrator's celebration of electricity seems as utopic as Mumford's. He asserts that "the truth is the light and light is the truth" (7), leaving readers to meditate upon this interesting but unexplained faith in electrical light.

When we consider the "battle royal" scene alongside this preview of the novel's conclusion, several interpretations of the scene become simultaneously available. In the prologue, the protagonist had invited readers to look for him in "the great American tradition of tinkers," alongside "Ford, Edison, and Franklin" (7). By invoking those American inventors and then depicting the cruel applications of modern inventions, these chapters reveal that electrification has not improved American life as General Electric or the Rural Electrification Administration had claimed it would. The abuses of human rights that pervade the "battle royal" scene suggest that the inventions of Invisible Man's purported kinsmen have not changed society as dramatically as most Americans at the time had been led to believe. Especially because Franklin and Edison were best remembered for their work with electricity, the use of this energy in the "battle royal" raises questions about their mythologized legacy in American history.

At the same time, the prologue and the narrator's attempts to "contain" electricity also encourage readers to anticipate hints of this character's inventive genius. When readers encounter the "battle royal" scene after reading the tantalizing prologue, they suspect that these white patriarchs will not be able to keep their social and electrical power to themselves for long. This anticipated outcome influences the way novel readers approach the "battle royal" scene. By juxtaposing these dramatically different accounts of electricity, Ellison insinuates that readers should understand the oppressive applications of this energy—*and* that they should realize that the oppressed person's relationship to these inventions is not wholly passive. Although electricity was used to enforce racial hierarchies historically and in the novel, overemphasizing this oppressive application "strips black people of technological agency," as the historian of technology Rayvon Fouché might argue.[25] The protagonist doesn't request readers' sympathy in these chapters; he demands their attention. Invisible Man wants readers to acknowledge his humanity. And to the character, that means recognizing how he finds ways to be creative within the intensely constrained setting of Jim Crow America.

While the "battle royal" scene emphasizes oppression and the prologue accentuates subversion, these scenes nullify the question of whether electricity is progressive when we read them side by side. Instead, they raise the alternative question: what has enabled electricity to signify progress in the face of flagrant social injustice? By shifting the conversation in this manner, Ellison conjures Mumford once again. In *Technics and Civilization*, Mumford argues that, in order to understand an invention, "[n]ot merely must one explain the existence of the new mechanical instruments: one must explain the culture that was ready to use them and profit by them so extensively" (4). In the "battle royal" scene, Ellison demonstrates how one relevant social group—wealthy, white, southern men—were prepared to use neotechnic inventions for their paleotechnic ends. Herein lies Ellison's technological humanist critique: until these powerful characters reframe their understanding of progress, electricity will continue to reinforce old social hierarchies rather than promote beneficial social change.

RESISTING RHETORICAL CLOSURE IN ELLISON'S "TRUEBLOOD" SCENE

In the subsequent chapter of *Invisible Man*, Ellison complicates this already sophisticated depiction of social and electrical power. He begins by placing this energy in a space that seems disjointed from the smoky hotel parlor in which the "battle royal" takes place: the campus of a historically black college that closely resembles the Tuskegee Institute. This scene opens with a reflective Invisible Man recalling his college days. He can still hear "the black powerhouse with its engines droning earth-shaking rhythms in the dark." Here Ellison re-imagines the power generator—an icon deeply associated with literal and metaphorical illumination—as an invisible, sonic force. "Black" and "dark," this dynamo sheds no light (34).[26] Its unseen rhythms resonate with the proto-Afro-futurist phonographic play of the novel's prologue.

This powerhouse seems distinctive from the white-owned electric circuits of the "battle royal" scene. Is this the "technology which promises [African-American people] release from the brutalizing effects of over three hundred years of racism and European domination" that Ellison hinted at in the interview "Some Questions and Some Answers"? Ellison invokes a parallel between his prologue and his second chapter as an enticing possibility. Ultimately, however, this dynamo fails to realize its potential. Throughout the scene, the narrator slowly unveils that white power actually motivates the school's machinery. Every decision made in the institution complies with the wishes of white multimillionaire trustees. In this context, the university's humming dynamo becomes an empty signifier. The campus doesn't have its own power. It generates electricity only so as to comport with the expectations of white donors, who want to be able to measure the progress of the campus according to their own standards of modernization.[27]

If electricity merely represents progress and doesn't actually promote positive change, then the notion of electrical progress resembles what SCOT scholars call "rhetorical closure" in that it uses language to convince relevant social groups that a socio-technical problem has been solved. And if rhetoric alone had created the impression that electricity promoted progress, then any crafty wordsmith could undermine that idea. In fact, Ellison introduces such a character in this chapter, shortly after the brief college interlude described above. The destabilizing figure is Trueblood, a poor, disgraced sharecropper who notoriously raped his own daughter in his sleep. Trueblood can't access electrical power in actuality; the dynamo that the narrator remembers had been designed strictly for campus use. Nonetheless, Trueblood weaves the school's generator into his narrative about his accidental incestuous experience. In the process, he undermines common assumptions about black sexuality and about electrical progress.

Follow Invisible Man as he meanders from the dynamo on his campus to the metaphoric dynamo of Trueblood's dream—his last stop before he visits the Golden Day. The protagonist has been a successful (or well-assimilated) college student, and for that he has been given the honor of chauffeuring a white multimillionaire. He "half-consciously" drives the trustee, Mr. Norton, to Trueblood's cabin on the outskirts of his campus. To the black university students, Trueblood represents the stereotype against which they resent being measured. The students and administrators fancy themselves modern; they disparage Trueblood—even before his "disgrace" (46–47)—because they consider him "primitive" (52).

Trueblood understands how the "biggity school folks" perceive him. He knows that they want to send him away to demonstrate their disidentification with him. They don't want to be compared to the farmer; they don't want to help or understand him, either. Trueblood resists their attempt to marginalize him by integrating the campus' dynamo into his rape dream—at least when he tells his story to a college student and to the wealthy white trustee. As the literary scholar Houston A. Baker argues (1983, 831), Trueblood "has thoroughly rehearsed his tale" and "has carefully refined his knowledge of his audience." Readers don't know if the raconteur uses the same electrical imagery in every retelling, or if he invokes the power plant to signify to this particular audience of college affiliates who have overinvested in the notion of "modern progress" that this artifact commonly signified.

The story that Trueblood shares with Mr. Norton and Invisible Man begins with a quest to find "fat meat." The journey leads him to a vaginal room. He continues: "At first I couldn't git the door open, it had some kinda crinkly stuff like steel wool on the facing. But I gits it open and gits inside and it's hot and dark in there. I goes up a dark tunnel, up near where the machinery is making all that noise and heat. It's like the power plant they got up to the school." According to this dream logic, Trueblood's daughter's vagina is "like" the college's power plant. In keeping with this symbolic economy, he remembers his orgasm "like a great big electric light in my eyes" (58–59).

Trueblood's story is part tragic, part trickster. Scholars such as Baker, Thomas Marvin, and Shanna Greene Benjamin have agreed that Ellison encodes this farmer's dream as a "tall tale."[28] These scholars perceptively recognize that the superficially powerless Trueblood has the power to "signify," in the sense in which Henry Louis Gates Jr. (1989) uses the word. Although white characters apprehend Trueblood as a reaffirming example of black degeneracy, and other black characters show disdain for him as an embarrassment to their race, he resists these narratives about himself by broadcasting a dream that "says nothing definitive about black sexual desire" (Johns 2007, 239). In addition to challenging assumptions about his atavism, Trueblood's narrative defiantly undermines the logic that codes electricity as modern, rational, and progressive.

In a compelling study of *Invisible Man*'s electrical imagery, Douglas Ford (1999, 898) reads this scene in terms of its binaries, arguing that Trueblood "depicts himself as essentially powerless, the scalded victim of an electricity that designates a boundary between rich and poor, white and black" and that "[a]t the same time, electrical images punctuate his discourse as untapped, unrecognized forms of energy in itself, forms that use verbal play to open the 'closed' circuitry of white power structures" (898). Ford divides the content and form of this passage. He claims that the content of Trueblood's story emphasizes the disparity between the haves and have-nots, while the form inverts that power dynamic. According to Ford, Trueblood's ability to signify resonates with the novel's prologue. Like Invisible Man at the beginning and end of the novel, Trueblood taps into figurative power that was never meant to belong to him.

Ford's account of "verbal play" expands productively on Gates's understanding of signifying, but his argument also dulls the affect of the scene. The Trueblood scene is unsettling, and any persuasive reading must account for that fact. This storyteller manipulates his audience profitably: Mr. Norton gives him a hundred dollars before leaving the cabin. But the pregnancy of Trueblood's daughter attests that there might be some truth to his story. Since the farmer's signifying cannot erase the themes of incest and rape, this scene resounds with the scene in *Martin Eden* with which I closed chapter 4. Like Jack London, Ralph Ellison places a scene of human profanity alongside an icon of modernity. This juxtaposition challenges the presumed correlation between electricity and social progress. Clearly, the development of electrical power systems has not improved Trueblood's life—or his daughter's life—in any way.

The Trueblood dream sequence is also disturbing for its familiarity. Electricity had long been sexualized in the American cultural imagination. In my introduction, I identified this tradition in the "venus electrificata" experiments and in Henry Adams's essay "The Dynamo and the Virgin." Ellison also hinted at the convergence between electrical power, white power, and sexism in the "battle royal" scene: while Invisible Man is being shocked by the electrified rug, the white patriarchs toss around a nameless "magnificent blonde—stark naked." And these are only three examples of a pervasive iconography. Electricity was consistently fashioned as a feminine force that was "tamed" by masculinized scientific expertise.[29] Such visual rhetoric proliferated in books and journals, in the stained glass windows of world's fairs, and even in live performances of young girls carrying or wearing electrical lights.[30] The Trueblood scene invokes these promotional depictions by translating the vague image of electricity as a feminine body into a narrative that associates electricity with sexual taboos. In the process, this scene draws attention to the power dynamics that already underwrite the desire for electrical control in the American cultural imagination.

While satirizing pervasive fantasies of electrical control, the Trueblood encounter also raises questions about Invisible Man's fixation on electrical power. It hints that electricity is not a neutral energy source that reflects the politics of its user. Electricity may seem destructive in the "battle royal" when it is used as a mechanism of oppression, and it may seem constructive in the college or in Invisible Man's apartment when it is used to demonstrate black technological agency. But regardless of who controls this energy and to what ends, it still has politics because Invisible Man has been led to believe that electricity *is* modern. The correlation between electricity and modernity has taught him to scorn folks like Trueblood and to respect white inventors such as Franklin and Edison.

The Trueblood scene troubles the protagonist's predilection for the bright, electrified, "modern" world. Although Trueblood is not a role model by any means, he has something to offer that Invisible Man cannot access: the disgraced farmer's face seems to transmit a "message" that Invisible Man "could not perceive" (51). This scene of failed communication hints that Invisible Man's value system is inherently flawed because it can't assimilate the lessons of his elders. Invisible Man's failure to understand Trueblood demonstrates negatively what Mumford had argued affirmatively: that only a person who can learn from narrative, from modern inventiveness, *and* from history can succeed in the twentieth century. As Ellison argues in his introduction to *Invisible Man*, "what is commonly assumed to be past history has actually as much a part of the living present as William Faulkner insisted." By conveying this point through the character Trueblood, the novelist modifies Mumford's message. Ellison implies that Americans—in addition to learning from the most "usable" elements of their history—must learn from the aspects of their past and their present that they are ashamed of.[31] This implied historiographical critique becomes more apparent as Invisible Man and Mr. Norton prepare to visit the "Golden Day," the bar named for Mumford's romantic whitewashing of history.

THE GOLDEN DAY

At the close of the Trueblood scene, the white trustee, Mr. Norton, suffers an emotional shock. The narrator implies that this character isn't disgusted by the tale of incest but rather is envious of it. Norton apparently harbored sexual feelings for his own late daughter. He asks his chauffeur, the nameless protagonist, for a stimulant. Invisible Man decides to take him to the Golden Day in search of whiskey. As chapter 3 opens, the two characters are making their way to the disreputable bar. Chapter 3 includes less electrical imagery than the previous two chapters and the prologue, but it displays Ellison's direct engagement with Mumford.

The literary scholar Alan Nadel has devoted an entire chapter to unpacking the "Golden Day" allusion in *Invisible Man*. As he argues, the frequent repetition of the bar's name in this chapter draws attention to the apparent citation of Mumford. When readers compare Mumford's book to Ellison's gambling parlor of the same name, they wonder what (if anything) these two Golden Days have in common and where Ellison's critique of Mumford will come into play (Nadel 1991, 95). More baffling still, Mumford used the title *The Golden Day* to describe a particular moment in history. As such, it remains unclear whether Ellison's "Golden Day" criticizes the book itself or the historical moment that Mumford and like-minded Americans over-idealized.

Lewis Mumford's *Golden Day* is a wistful work of literary criticism. It surveys American and European literary and cultural history from the perspective of an aesthete who feels entitled to enjoy literature about broad, humanist questions. In the slim volume, Mumford discusses a coterie of elite pre–Civil War and post–Civil War writers. Although he admits that Jack London and Theodore Dreiser stood out as powerful writers in their own time, he ultimately claims that that the *fin de siècle* generation had a limited sense of the possible. Whereas late-nineteenth-century writers "idealized the real" (Mumford 1926, 81), romantics strove for the improbable. Mumford argues that a revived romanticism is required to move American intellectuals out of their oppressively pragmatic rut. Mumford advances this argument by mapping his favorite romantic writers onto an extended metaphor of a day. He describes Emerson's writing as a dawn of the individualist spirit in American letters, positions Whitman at the noon-like climax of Golden-Day values, correlates Hawthorne with the twilight, and situates Melville at the darkening close of this figurative heyday of American individualism. This extended metaphor requires some rhetorical acrobatics. As Nadel (1991, 88) notes, in order to privilege Whitman's admirable blend of humanism and scientific curiosity, Mumford positions Whitman before the gloomier Melville—even though *Leaves of Grass* (1855) was published after *Moby-Dick* (1851). And this is not the only data point that Mumford manipulates. His distortions and omissions are legion.[32]

As I indicated above, Mumford's account of the Civil War was particularly contentious. Consider this passage from *The Golden Day*, which reveals both the type of thinking that Ellison disparaged and the embryonic ideas that would mature into *Technics and Civilization*:

> The Civil War cut a white gash through the history of the country; it dramatized in a stroke the changes that had begun to take place during the preceding twenty or thirty years. On one side lay the Golden Day, the period of an Elizabethan daring on the sea, of a well-balanced adjustment of farm and factory in the East, of a thriving regional culture, operating through the lecture-lyceum and the provincial college; an age in which the American mind had flourished and had begun to find itself. When the curtain rose on the post-bellum scene, this old America was for practical purposes demolished: industrialism had entered overnight, had transformed the practices of agriculture, had encouraged a mad exploitation of mineral oil, natural gas, and coal,

and had made the unscrupulous master of finance, fat with war-profits, the central figure of the situation. All the crude practices of British paleotechnic industry appeared on the new scene without relief or mitigation. (79)

Although Mumford uses the term *paleotechnic* here, he doesn't flesh out a schema of technical phases at this point. *The Golden Day* doesn't idealize neotechnics or electricity in the way that *Technics and Civilization* would, eight years later. Nonetheless, the stakes of both books are similar. Both urge readers to overthrow their instrumentalist ideologies in order "to conceive a new world" (144). In *Technics and Civilization*, Mumford encourages his readers to imagine a new social and technological future, illuminated and beautified by electricity. In *The Golden Day*, he promotes the new world he imagined with literary and cultural nostalgia.

While the above passage serves the rhetorical purpose of glorifying a historical era and criticizing the writer's current moment, it also muddles Mumford's day metaphor. It awkwardly combines the Elizabethan seafaring age, the dawn of industrialization, and the rise of the university system as if they constituted one coherent historical moment "in which the American mind had flourished." It then reduces the Civil War into a nefarious turning point. When Mumford's day draws to a close, "paleotechnic industry" creeps in, as if "overnight." From that point on, Mumford argues, mechanical slavery takes the place of human slavery.

This notion that all Americans—even artists—had become "slaves" to the machine is the central anxiety that motivates Mumford's early work. According to *The Golden Day*, after the Civil War (or, as Mumford called it, the "struggle between two forms of servitude") ended, the depletion of the human spirit began. Mumford claims that the "people who *did* believe in the triumphs of the land-pioneer and the industry-pioneer ... might quote statistics till the cows came home: they had only to look around them to discover that, humanly speaking, they were in the midst of a dirty mess. Machines got on. ... But men and women—they somehow did not reflect on these great triumphs by an equivalent gain of beauty and wisdom" (91–92). The generalizing presumption that all men and women were worse off after the war presumes that slavery had been more palatable than the rise of machines. Mumford's appeal seems compassionless from Ellison's perspective; the nonfiction writer allows his desire for "beauty and wisdom" to eclipse any empathy for the enslaved person's desire for freedom.

While Ellison takes issue with this revisionist account of the Civil War, he still finds value in *The Golden Day*. In fact, even when Ellison disagrees with Mumford's reasoning, he often agrees with Mumford's literary assessments. Both Mumford and Ellison concur that early-nineteenth-century writers were more likely to write about how to "live a whole human life" (Mumford 1926, 142) than recent writers were. To Mumford, writers indicated their investment in the whole of human life by "welding together the interests which science represented, and those which, through the accidents of its historic development, science denied" (141). To Ellison (1964), writers revealed their commitment to this theme by acknowledging that black people are fully human, rather than abstract symbols or stereotypes.

With these resonances and dissonances in mind, we can more thoroughly comprehend what Ellison means when he writes of the "Golden Day." That suggestively named establishment seems to house every marginalized figure in American culture. Some are veterans, whose service has been forgotten by a culture that fails to treat or acknowledge their trauma. Some are failed professionals who have gone insane. Some are deviant women. These characters signify the people whose lives have not been improved by the various inventions that seemed to connote progress in mainstream white American culture. By writing this cast of characters into a bar called the "Golden Day," Ellison re-inscribes the various types of people who Mumford erased from his romantic emplotment of American history.[33] In this context, the "Golden Day" allusion sheds light on the Trueblood scene that precedes it. Both of these scenes challenge writers like Mumford to stop erasing the aspects of history that shame or confound them.

Ellison's shell-shocked veterans specifically undermine Mumford's description of war. Although these characters are veterans of World War I and not of the Civil War,[34] they nonetheless remind readers that wars are not just symbolic turning points. Wars are traumatic, and their casualties exceed the number of people who died in battle.[35] Ellison's version of The Golden Day implies that the people who fought and suffered in any war continue to matter even after the war has ended. Their dashed hopes are as important as the fettered hopes of the humanist in a machine-obsessed age.

Wright (2005, 162) argues that this tension between Mumford and Ellison is fundamentally an infelicity of scale: "The fate of urban culture, the possibility of constructing a global technological utopia that was *not* technocratic, the hopes and frustrations of large-scale urban planning—all impelled Mumford toward a collectivizing, macrocosmic vision." Meanwhile, Ellison was preoccupied with the affects of the system on the individual's personality and psychology. The fact that these authors can attend to micro- and macro-scale perspectives with consonant electrical imagery suggests that this energy still symbolized the nexus between individual and system. This difference in perspective also affects the conclusions each writer can draw. Whereas Mumford writes about technics and civilization, Ellison writes about technics and personhood. Thus, Ellison is able to resist the tendency toward generalization that becomes Mumford's signature strength and greatest pitfall.

The form of the novel also allows Ellison to criticize Mumford obliquely. In fact, the discrepancy between the fiction writer's and the nonfiction writer's interpretations of technics and society become most apparent when Ellison introduces an interesting but unreliable character who speaks in Mumfordian prose. One Golden Day patron, who the narrator interchangeably calls "the doctor" and "the vet," particularly ventriloquizes Mumford. Ellison doesn't explicitly describe this character's race or his physical body, but he implies that the man is non-white and likely black when Invisible Man marvels to hear "him talk as he had to a white man" (93). The former doctor, now a patient, gains ethical appeal by correctly diagnosing Norton's condition. "Your diagnosis," the white trustee tells the man, "is exactly that of my specialist ... and I went to several fine physicians before one could diagnose it" (90).

At Norton's urging, the doctor recounts his history. Insofar as this conversation takes place in a venue called The Golden Day—and as it is triangulated by a character (Hester) who evokes one of the major figures of *The Golden Day* (Hawthorne)—the dialogue compels readers to acknowledge the allusion to Mumford. Indeed, the doctor's life story parallels Mumford's tragic interpretation of American history: this character had become an expert in a scientific and technical field, but he still felt that some aspect of his spiritual and emotional life was lacking. He seems to agree with Mumford that a scientific life is not enough for the human spirit. He explains that he had been a successful brain surgeon in France, but that he returned to United States for one reason: "Nostalgia" (92).

Like Mumford, this character tries to redress that sense of lack by succumbing to his longing for times past. But readers soon learn that the doctor's nostalgia is more dangerous than the white writer's; non-white Americans do not have the luxury of romanticizing America's past. When the doctor returned to America and continued acting as a surgeon, "ten men in masks drove [him] out from the city at midnight and beat [him] with whips for saving a human life" (93). This modern man cannot find "human dignity" (92) in his work—but not because of the inhumane logic of machines, as Mumford predicts. The doctor's work fails to bolster his human dignity because of the inhumane logic of white supremacy.

This veteran invokes Mumford more forcefully as he shifts from telling his own story to addressing Mr. Norton and the nameless protagonist. He calls the protagonist "the mechanical man" and an "automaton" because "he [the protagonist] believes in that great false wisdom taught slaves and pragmatists alike, that white is right" (94–95). In Mumford's body of work, *slave*, *automaton*, and *mechanical* describe how machines have hindered the Transcendentalist aspiration for greatness. By using Mumford's terms to describe the arrested development of the novel's protagonist, Ellison draws an analogy between racism and machine-age utilitarianism. According to Mumford, white people might become automata by finding themselves trapped "in a sort of mechanical dream" (152). According to Ellison, black people have been trained to act as automata all their lives—regardless of his relationship with new inventions—because the white majority denies their humanness and demands from them a scripted set of automatic behaviors. At this point in the novel, Invisible Man is a mechanical man because he acquiesces to these dehumanizing expectations.

ELECTRICITY AND HUMANISM IN THE "OUT OF THE HOSPITAL" SCENE

Ellison examines the aforementioned theme repeatedly throughout his work. In a nonfiction essay for the Federal Writer's Project, "The Way It Is" (1942), he describes how employers used unequal access to technical training in order to reinforce white supremacist hiring practices.[36] In the short story "Flying Home" (1944a), he explores how a nameless black pilot's overestimation of his airplane and devaluation of black folk knowledge could result in a depleted sense of self. In another short story, "King of the Bingo Game" (1944b), he constructs yet another nameless black protagonist who has an identity crisis as he presses and refuses to release an electric button. In the novel *Three Days Before the Shooting* (published posthumously in 2010), Ellison portrays how a black character could hope to convey his wealth and success by buying a car, and he shows how this expensive acquisition would devalue the car before it could lend value to the black man's life. Ellison consistently dramatizes the personal tragedies that follow from a misguided faith in technology. Dedicated to ambivalence, he doesn't argue this point; he impresses it upon the page like grooves on a record. He invites readers to listen for the resonances and dissonances between racism and Mumford's mechanical-existential crisis.

Among his many narratives about race, technology, and identity, the "Out of the Hospital" scene in *Invisible Man* stands out as a particularly powerful example for close reading. This scene blends humanistic and electrical imagery more explicitly than Ellison's other fiction. It also serves as a fulcrum. It is Invisible Man's tipping point; it changes his identity and his politics. Before his hospitalization, he still believed in the idea of progress. Afterward, he feels drawn toward anarchy.[37] Furthermore, Ellison penned multiple versions of this episode, and each variation offers insight into the different orientations the writer took toward technology in general and electricity in particular.

Alternate drafts of the "Out of the Hospital" scene have already attracted scholarly consideration. According to Barbara Foley (2010, 207–208), this section of the manuscript consumed Ellison's attention more than any other. Indeed, Ellison demonstrated his investment in the original manuscript's alternate story line when he revisited and published it as a self-contained short story, "Out of the Hospital and Under the Bar" (1963). Scholars who have studied that story, including Shanna Greene Benjamin, Melvin Dixon, and Claudia Tate, have suggested that we should not limit our understanding of *Invisible Man*'s hospital scene to the published chapter.[38] Although I follow their lead in that respect, it is worth noting that their interest in the "Out of the Hospital" variant differs from my own. These previous scholars recuperated the alternate version because it expands the agency of Mary Rambo, a folk character who plays a limited but enticing role in *Invisible Man*. I focus primarily on the electro-shock scene of the manuscript variant—a scene that doesn't appear in the published short story because it precedes Mary's entrance to the scene.

The version that Ellison included in the published novel begins shortly after Invisible Man is dismissed from his university for bringing Mr. Norton to Trueblood's cabin and to The Golden Day—sites of shame that the university administrators had hoped to keep hidden from their investors. In intervening scenes, the protagonist had traveled north and procured a job at the Liberty Paints Factory. But his employment had ended abruptly with a furnace explosion, and the hospital scene begins when the narrator awakens after that accident. In this vignette, electricity once again plays an ambiguous role. White doctors force the protagonist to undergo electro-shock therapy, but this act of violence imbues the patient with an unexpected sense of kinship with his doctors.[39] It is the incongruous juxtaposition of that violence and sense of filiation that concerns me here.

Ellison stages the hospital scene as equal parts horror and science fiction. The narrator awakens in his hospital bed to find seemingly nonhuman doctors peering down at him. They have been rendered alien-like by technological prosthetics: "I was sitting in a cold, white, rigid chair and a man was looking at me out of a bright third eye that glowed from the center of his forehead" (231). Reciprocally, the doctors see the protagonist as nonhuman. They treat him as a subject of scientific interest rather than a patient. The protagonist overhears, first, a debate about whether to release him or to "shoot him up for an X-ray," and, later, the sound of a "machine [beginning] to hum." The decision was made without consultation.

Once the machine warms up, Invisible Man is "pounded between crushing electrical pressures; pumped between live electrodes like an accordion between a player's hands" and he realizes through the intense, immediate pain "that [his] head was encircled by a piece of cold metal like the iron cap worn by the occupant of an electric chair." Building on a theme that Ellison explored in his earlier short fiction, this scene describes a nameless man who submits as white men assert their control over his physical body and life. The quick allusion to the "electric chair" heightens this resonance, alluding to the long history of legal and illegal executions of black individuals. Much of these foregoing details already appeared in Ellison's manuscript chapter. He apparently crafted the images of the doctor's uncanny third eye and of the electric chair cap early in his writing process.

Later in this passage, Ellison describes Invisible Man's electric shock synesthetically and metonymically. His protagonist experiences the electrical current as a blend of x rays, electricity, radio waves, phonography, and, arguably, even telegraphy (in that his doctors try to communicate with him in succinct bursts).[40] In both drafts, Ellison constructs hybrid images in order to convey the multifariousness of the electrical current, thereby building on the literary tradition that I examined in previous chapters. According to Invisible Man, an electrical current is painful, beautiful, musical, and somehow feminine in that he hears a woman's voice as the charge passes through him (234–235).

After the first currents of electricity pulse through the protagonist, he forgets his own name—a crucial element of the narrative that had been unexplained up to this point. He recalls: "I tried to remember how I'd gotten here, but nothing came. My mind was blank, as though I had just begun to live.... I seemed to go away" (233). This dual experience of fading and beginning is another hybrid image. It plays on the paradoxically fatal and life-giving qualities of the electric charge. Strapped in this medical electric chair, the protagonist embodies the paradox of Schrödinger's cat: he is lost in a liminal state that is both alive and dead while he awaits medical observation in a nickel-and-glass display case.

Not even the experts who manipulate the protagonist's fate are certain of what will become of their patient/victim. One doctor purports that new inventions will somehow help the patient, claiming "my little gadget will solve everything" (235). His fascination with his own equipment evokes the trope of the mad scientist. More concerned with his new "gadget" than with his patient, this doctor never defines the problem he hopes his invention will "solve." In contrast, the other attending physician is convinced that the fruits of electrical progress cannot apply to this nonwhite patient: "I think I still prefer surgery, and in this case especially, with this, uh ... background, I'm not so sure that I don't believe in the effectiveness of simple prayer" (235). Assuming that the patient is "primitive," this doctor contends that Invisible Man should be treated only by "primitive" means, such as invasive surgery or prayer.

In this scene, Ellison demonstrates the stakes of the modern/primitive binary from a different perspective. Just as Invisible Man had interpreted his campus dynamo as modern and Trueblood as primitive, these doctors interpret their equipment as modern and their patient as primitive. The two physicians debate the primacy of their dichotomous categories: Should the presumed modernness of the "gadget" guarantee its usefulness? Should the presumed primitiveness of the patient invalidate the modern artifact's utility? Although this debate registers some tension between racism and technological determinism, the narrator ultimately correlates the two ideologies because they yield the same dehumanizing result. Either the hero is denied medical care because of one doctor's racial bias, or he is involuntarily made the subject of an experiment because of another's General Electric–like "electrical consciousness."

In this scene the latter bias wins out, but the experiment has an unintended effect. First the protagonist loses a sense of his own body's boundaries; then he identifies with his doctors: "I could see smut in one doctor's nose; a nurse had two flabby chins. Other faces came up, their mouths working with soundless fury. But we are all human, I thought, wondering what I meant" (239). In the published version alone, the protagonist's experience of unboundedness suddenly allows him to perceive himself and his doctors in a different light.

Previous scholars have interpreted the published scene as a narrative about Invisible Man's subversive cyborg identity.[41] But when we read Ellison in conversation with Mumford, the humanist undertones of this scene appear more prominent than the cyborg subtheme. In this context, the gadget-obsessed doctor seems like one of Mumford's victims of machine enslavement: this character is entirely instrumentalist and devoid of sympathy, in accordance with societal expectations about scientific expertise. And Ellison, like Mumford before him, challenges the presumed superiority of these utilitarian values by comparing them with a romantic alternative. When Invisible Man loses track of his body's boundaries, he doesn't question his innate humanness from his electrified hospital bed. He reasserts it.[42] By drawing attention to the humanness that this character can see and that his doctors cannot, Ellison invites readers to stop seeing Invisible Man as a sympathetic victim. Notably, this humanist insight comes from within the machine. In this scene, Ellison doesn't set humanism against mechanism; he sets technological humanism against inhumane utilitarianism.

THE MANUSCRIPT VARIANT

When Invisible Man thinks "But we are all human," his epiphany is exciting and discomfiting for character and reader alike. Even Invisible Man cannot understand what he means when he identifies with the ugly fleshiness of his doctors. This feeling of shared humanity resists facile classification. It is transcendent and violently oppressive. It is also unidirectional. There are no indications that the doctors see any fragment of their own humanness in the body of their patient. The asymmetry of this humanistic moment registers the fact that Ellison touched upon in "Some Questions and Some Answers": science and technology reproduce the uneven power relationship that they were developed within. This doctor's new invention, ostensibly designed to heal, becomes yet another instrument for exerting white power over a non-white person; it doesn't "solve everything," after all. Still, neither this device nor the shame of being experimented upon can extinguish Invisible Man's faith that humanness matters. For this electric instant, the protagonist meditates on the profound essence that makes human life worth protecting. His affirmative embrace of his and his doctors' humanity seems more compelling than either doctor's electrically or racially mediated account of the situation.

This sequence was a late addition to the novel. Ellison's manuscript chapter excludes the three-line realization that culminates in the line "But we are all human." The surrounding conversation is otherwise remarkably similar. In Ellison's manuscript chapter, the theme of humanism effervesces a little later in the scene. This unpublished version begins with an incongruous juxtaposition between sterile machine and human bodily function: "What were the instruments radiating? Then in my effort to hold my breath I swollored [sic] it and allowed a lazy burp to bloom up and barely suppressed it. It was as though someone had yelled 'rape!' The nurse look horrofied [sic]. Hell broke loose. The space above the case filled with alarmed faces. Frantic tests were begun on all the equipment. The doctor-inventor rushed in" (Ellison ca. 1950, 270).

This scene parodies the "doctor-inventor's" expertise. Like the published scene, the manuscript variant is also Mumfordian in that it pokes fun at the medical professionals who are so entranced by machines and so far removed from their patient's humanity that they misconstrue a burp as a serious mechanical error. No one in the room can imagine that this patient has any natural bodily functions, much less any agency. In this sense, this scene satirizes the white American tendency to over-invest in "gadgets" and to overlook black humanness. Ellison once again yokes Mumfordian fears about mechanical enslavement together with his own concerns about racism.

Eventually in the manuscript version, Invisible Man burps a second time. This time a doctor notices: "Him, right there. He's the cause!" After some reflection, Invisible Man muses: "I actually felt thankful to him for discovering the burp. At least he seemed to hold me responsible while the others regarded the incident simply as an interruption to the smooth function of the machine. He, for a moment at least, was aware of a/process originating within me. Thanks, humanist" (271). This reference to humanism lacks the subtle profundity of Ellison's published chapter. The cultural critique of the manuscript version is explicit, maybe even heavy handed. Throughout this instantiation of the scene, Ellison ridicules the machine-obsessed experts who cannot recognize flatulence. His response—"Thanks, humanist"—seems glib. Here, the so-called humanist doesn't acknowledge anyone's humanity. The doctor merely recognizes that the patient's body has basic physiological functions.

After these interludes, the two versions of this scene further diverge. In the published novel, the doctors tell Invisible Man that his treatment was complete. They suddenly and inexplicably let him go. In the manuscript, Invisible Man escapes the machine without the doctors' approval. In this version, Ellison introduces the character Mary Rambo one chapter earlier than he does in the novel, and he gives her the rebellious power to liberate Invisible Man from his glass case. This alternative outcome begins with Mary's entrance on the scene: "Out of the swiftly receding blur she grinned down with bright black eyes. Rays of light glinted off coppery teeth that held a frayed toothpick" (Ellison ca. 1950, 260). At first, the adjective "coppery" seems only to set Mary apart from the doctors, who are hygienic, starched, white. But as Mary reveals her insurrectionary understanding of the electric machine that holds Invisible Man, she comes to resemble copper wire: she invisibly moves power across the electrified white, hospital. In the manuscript and in its revision, "Out of the Hospital and Under the Bar," Mary decides what circuits to close and to open.

In this alternate story line, Invisible Man initially fears Mary's help. When she begins to pull at his case, he wonders "What if she was opening the case too soon? Before my treatment was complete? Suppose she turned the wrong bolts and set the machinery in motion?" (Ellison ca.1950, 278). The protagonist hesitates in fear; he has internalized the hierarchical structure of white modernity and black primitiveness, of male expertise and female passivity. He explicitly reflects: "For though I wanted release, I was frightened lest it should come through this ignorant, unscientific old woman" (278). Presuming that Mary could not understand the sophisticated machine that encases him, he worries than this "unscientific old woman" might inadvertently electrocute him. But Mary opens the case, turns off the light, and allows Invisible Man to unplug himself.

To prepare Invisible Man for escape, Mary gives him a pork sandwich and a cluster of dried leaves (a folk cure). As Benjamin notes (2014, 135), these gifts lend to Mary Rambo's characterization as "the fundamental repository of black self-sufficiency and cultural independence that the protagonist tragically lacks." Mary enables Invisible Man to perform his own rebirth: with the nourishment of the pork sandwich and the inexplicable magic of the "root," he casts off his electric-umbilical cord and escapes into the sewers, only to emerge at Mary's home and into the next scene of the novel.

Fouché's (2006) concept of "black vernacular creativity" offers a useful vocabulary for understanding the differences between these two variations of the hospital scene. The manuscript version doesn't feature any "vernacular creativity." It depicts electricity—or at least the singular medical invention of this scene—as a lever of white supremacy. This apparatus is one material way that white experts exert control over a helpless black man. This plot defines the relationship between electricity and African Americans as unilaterally antagonistic. Had this version been included in the novel, Invisible Man's use of incandescent light in the prologue might appear as a form of capitulation rather than a form of subterfuge.

In the published chapter, Invisible Man transforms a machine that was designed to dehumanize into a machine that reaffirms his humanity. In so doing, he performs what Fouché calls *reconception*: this character acts in a way that "transgresses that technology's designed function and dominant meaning" (Fouché 2006, 642). If we recall the accusation from the

"Golden Day" scene—that the protagonist is a "mechanical man"—then this scene marks a moment of transformation. Invisible Man may remain "mechanical" in the manner that Mumford or "the doctor" implied: he lacks a fully developed sense of his own humanity. But this mechanical man has at least been modified with a signal converter. White authorities—from the "battle royal" patriarchs through these hospital doctors—had tried to hard-wire this character, to ensure that he would react in scripted ways to specific social and electrical stimuli. In this scene, Invisible Man takes the same input, but he yields a different output than was anticipated. He absorbs electrical currents and racism, but he translates this oppressive signal into a life-affirming experience. Thus, although the character cannot control electricity or his own body in either the manuscript version or the published version of this scene, the latter rendition offers the character more power to "tinker" subversively with inventions that were designed to degrade him.

These two alternate endings dramatically change Ellison's depiction of electricity. The manuscript version of this story line promotes an anti-establishment orientation toward "electrical progress." When Mary pits her folk knowledge against the modern expertise of the hospital, she wins. Mary's success enables the protagonist, who once gave a scholarship-winning speech on progress, to learn that his own growth might involve looking backward to older forms of folk knowledge instead of looking forward to new gadgets. (Indeed, this manuscript variant might invite readers to wonder what the protagonist could have gained by communicating more effectively with a different folk character, Trueblood, several chapters earlier.) This story line offers a fantasy of rebellion—a fantasy in which "unplugging" becomes a real possibility. In the published version, Ellison offers a qualified, internalized form of resistance. Invisible Man lies passively in his glass case, but he doesn't allow his brain-altering electro-shock treatment to function in the way the doctors intended.

These endings coexist as two possible strategies for defying oppressive technological regimes. Juxtaposed, they shed light on Ellison's electrical imaginary. And, although these stories are about much more than electrical systems, they offer a usable history of electricity. When we read Ellison's manuscript today, Invisible Man's act of unplugging speaks to an enduring fantasy about breaking free from the systems that we habitually and often unthinkingly use. In fact, the verb "unplug" has become a dead metaphor for relaxation. But, when unplugging proves untenable, Ellison's published chapter reminds us of a small but feasible alternative: we can choose to make decisions about technological use that recognize our humanness and human connections, even when we feel powerless to initiate broader systemic change. The simple act of reframing what an invention means to us might open room for "interpretive flexibility."

WHO, ON THE LOWER FREQUENCIES, SPEAKS FOR YOU?

When compared against Mumford's utopian visions, Ellison's technological humanism may seem somewhat less grandiose. Mumford's early studies invite readers to develop a new understanding of technics and society that could stimulate human vitality; Ellison's *Invisible Man* encourages readers to hope that an abused and electrocuted black man might be set free from the mechanical, electrical, and social systems that would harm him. But this disparity doesn't imply that Ellison's account applies only to black Americans or that Mumford's has lost all of its value because of the limitations of his privileged perspective. It suggests, rather, that these already canonical and compelling texts gain poignancy when we read them together.

When we read Ellison in conversation with Mumford, it becomes more apparent that the novelist's African-American protagonists are not the only ones who repeatedly suffer at the hands of new inventions. His white characters are also victims of false consciousness: they have been taught to invest in a dehumanizing notion of progress. These racist and deterministic characters have been trained to build or buy new machines, when they might have been striving toward more fulfilling personal ambitions. By taking up these issues, Ellison layers an imaginative and intimate account of racialized power dynamics onto Mumford's macrocosmic model, leaving readers with a version of technological humanism that accounts for the perspectives Mumford omitted.

Ellison draws further attention to his Mumford-inspired subtheme in the epilogue to *Invisible Man* when he reminds readers that he was reading and thinking about the public intellectual as he was writing the novel. As his narrator concludes the story, he describes a recent run-in with Mr. Norton, the one-time trustee from the Trueblood scene, who is now old, feeble, and lost. Invisible Man tells the multimillionaire "Take any train; they all go to the Golden D—" (578). But Norton doesn't offer the narrator the opportunity to finish his thought. The old man slips away onto a passing train. This final mention of the Golden Day at the close of this 581-page novel lends emphasis to the Mumford citation. It indicates that, among the myriad allusions that the novelist made throughout this masterpiece, the nod to Mumford was still important enough to warrant reiteration in the final pages.

The narrator embellishes this explicit reference to Mumford by implicitly engaging with the public intellectual's philosophies in these final, summative pages. Reflecting on his life, Invisible Man adopts a neo-Transcendentalist subject position that is analogous to the one that Mumford promoted in *The Golden Day*. The narrator explains: "[L]ike almost everyone else in our country, I started out with my share of optimism. I believed in hard work and progress and action, but now, after first being 'for' society and then 'against' it, I assign myself no rank or any limit, and such an attitude is very much against the trend of the times. But my world has become one of infinite possibilities" (576). In this passage, Invisible Man chooses to act like a modern Emerson and to abandon his goal-oriented aspirations. He decides that any end—progressive or anarchic—would be existentially unfulfilling. By giving up the presumption that he should work toward some specific end, he has claimed to uncover "infinite possibilities."

Still, skeptical readers might wonder why they should believe that this character's possibilities are infinite, when he has confessed to hibernating underground as a social outsider. The narrator elucidates this question, but doesn't directly answer it, by going on to meditate upon his own humanity. This monologue demonstrates that the narrator's limited potentials in a segregated nation are not the "possibilities" that should define him:

> Perhaps that [by which he means his newfound orientation towards life] makes me a little bit as human as my grandfather. Once I thought my grandfather incapable of thoughts about humanity, but I was wrong. Why should an old slave use such a phrase as, "This and this or this has made me more human," as I did in my arena speech? Hell, he never had any doubts about his humanity—that was left to his 'free' offspring. He accepted his humanity just as he accepted the principle [of American democracy.] It was his, and the principle lives on in all its human and absurd diversity.

What would cause the narrator to question his own humanity if his grandfather experienced no such insecurity? Ellison leaves this question unanswered. By retaining this ambiguity, he implies that the protagonist's "infinite possibilities" map onto the countless ways that his life could hold meaning. Having thrown off a telic understanding of life, Invisible Man no longer has to work toward success by anyone's standards. He can redefine his own life's worth by communicating with the audience—by asking to be heard even if he cannot be seen.

Invisible Man's rhapsody to humanness echoes the epiphany that he experienced in the "Out of the Hospital" scene. As in that scene, he translates a painful experience into a reaffirmation of his own humanity. And, as in that scene, he anticipates that others will fail to recognize the shared humanness that he feels. He hurls an accusation at readers: "You won't believe in my invisibility and you'll fail to see how any principle that applies to you could apply to me. You'll fail to see it even though death waits for both of us if you don't." Yet, despite his fatalistic expectations, Invisible Man ends on a hopeful note. He closes the epilogue with the arresting line: "Who knows but that, on the lower frequencies, I speak for you?" This conclusion is striking for multiple reasons. It pays homage to Walt Whitman.[43] It poses the socially progressive claim that a black man could become the representative voice of his generation, even in the Jim Crow era. It achieves the aesthetic ambiguity to which Ellison aspired. And, when we read Ellison in conversation with Mumford, this passage also hints that Invisible Man has found a way to privilege his personhood over his fascination with technologies.

Ellison crafts this impression by way of elision. The "lower frequencies" he describes in his final lines conjure—but never actually describe—a radio. And the radio is not the only artifact Ellison invisibilizes here. This chapter returns readers to the scene of the prologue. But where the prefatory chapter pulsated with phonographic sound, the conclusion focuses only on the narrator's voice. And where the prologue depicted cascading rays of electric light, the epilogue is conspicuously lightless. It is not unelectrified; the narrator simply allows the 1,369 lights to go unmentioned. The human element now eclipses the technical, much as Mumford predicted that it would in the most utopian moments of *Technics and Civilization*. Where Ellison's protagonist once projected meaning onto electricity and other inventions, he now deemphasizes these inventions so that he may cultivate meaning in and for himself.

By opening and closing his novel in this way, Ellison narrativizes Mumford's romantic philosophy of technology. But the novelist simultaneously challenges Mumford's claim that this new subject position would be a panacea for the modern condition. As the narrator describes his journey toward self-re-definition, he explains that his change of heart was painful and difficult and shaped by the history of American slavery—even if his new way of seeing was ultimately preferable to his earlier illusions. Ultimately, Ellison agrees with Mumford that Americans should learn to appreciate their humanness over their modernness or inventiveness, but he also admits that this new epistemology will not solve every social or spiritual crisis.

Ellison's engagement with Mumford throws these moments from *Invisible Man* into relief. It helps readers appreciate why such a careful novelist would include so many rich depictions of electricity in a work that ultimately focuses on humanness above all else. At the same time, Ellison extends Mumford's social and technical vision. He offers historians of technology—or any reader who is interested in reconsidering her culture's utilitarian priorities—an intimate and artistic account of the different questions we might ask about the place of technology in American daily life. He invites readers to ask who we want to speak for us: the proponents of progress, who insist that new inventions will free us from figurative enslavement, or the person who is struggling to relearn what it means to be human in a chaotic and frightening and electrifying world?

CONCLUSION

Today, information commands our attention more frequently than electricity. The word *information* functions as a shorthand for conveying the networked character of the United States in the twenty-first century. We describe ourselves as living in an information age, we understand our bodies in terms of information and "genetic codes," and we re-interpret longstanding social inequalities in terms of the discrepant access we have to this information.[1] As I have shown, the concept *electricity* functioned in a similar manner during the late nineteenth and the early twentieth century. It metaphorically represented progress and democracy, although the asymmetrical illumination of cities re-inscribed social hierarchies by leaving poor areas unlit. Despite the tension between its progressive symbolism and its unequal implementation, electricity was a medium that enabled many daily routines and was an icon that remained conceptually and aesthetically alluring even as it became commonplace. Writers understood electrical power as a facet of American life that could provoke new ways of thinking about that life. In attempting to use electricity to think differently, they metaphorized this feature of the physical universe in imaginative and often incongruous ways.

In my introduction, I asked why writers took up these issues at all—why they projected meanings onto electrical systems. I suggested that the multifariousness of electricity was one answer to this question. Creative writing rests on the load-bearing line, and electricity allowed writers to signify multiple images simultaneously with an artful economy of phrasing. But there is another compelling answer to consider. Nineteenth-century and twentieth-century writers took up the subject of electricity because the newly accessible energy had become implicated with their construal of their own lives.

Adams, Twain, Gilman, Mumford, and Ellison each suggested that electricity had changed his or her understanding of human existence. And the reciprocal is also true: their sense of self—their concerns about being individuals in a time of system construction—shaped their understanding of this energy. They corroborate Steve Woolgar's claim that "discussions about technology—its capacity, what it can and cannot do, what it should and should *not do*—are the reverse side of the coin to debates on the capacity, ability, and moral entitlements of humans" (2012, 304).

This special relationship between bodies and electricity inspired new experiments in electrical science, in physiology, and in capital punishment. It also inspired poetry. Consider "Her Lips Are Copper Wire" from Jean Toomer's modernist pastiche *Cane* (1923). This poem can speak to the social shaping of electricity; it hints that we have inscribed our desires onto the electrified cityscape. It reciprocally reveals that role that electricity has played in shaping our conceptions of beauty, of city life, of bodies. Succinct, it is worth quoting in its entirety:

> whisper of yellow globes
> gleaming on lamp-posts that sway
> like bootleg licker drinkers in the fog
> and let your breath be moist against me
> like bright beads on yellow globes
> telephone the power-house
> that the main wires are insulate
> (her words play softly up and down
> dewy corridors of billboards)
> then with your tongue remove the tape
> and press your lips to mine
> till they are incandescent

This poem sings the sensuality of copper. Although the conductive metal appears only in the title, each stanza seems to follow the fine wire as it conveys power up lamp-posts, into telephone receivers, into the power-house, and beyond. As a whole, the poem registers the promiscuousness of this metal as it meanders unseen across the modern cityscape, from the objects we hold in our hands (like the telephone) to the artifacts (or people) against whom we press our lips. The copper is sheathed, "insulate." But beneath its thin, easily stripped "tape" lies the metal that experiences, invisibly, the excitement of electrons that we perceive as incandescence.

Perhaps this poem allegorizes a simple human kiss, as literary scholars have most often presumed. Perhaps it portrays a person who wants to feel an actual electric spark pass through him, like the curious power plant visitors I discussed in chapter 2. Perhaps "lips" are the metaphor, and Toomer is eroticizing and anthropomorphizing electricity itself. The poignancy of this poem lies precisely in this indeterminacy. "Her Lips Are Copper Wire" eroticizes the liminal, linguistic space where electricity and humanity become indistinguishable.

UNDERSTANDING ELECTRICITY AS A METONYM

Toomer's poem seems to share little in common with present-day conceptualizations of electricity. Indeed, the idea of electricity surging through a human body has lost some of its appeal, and these days we pay little heed to the wandering of copper wires. Acclimated to electricity, we react more considerably when the lights go out than when they beam on; the very idea of an un-electrified world is the stuff our dystopias are made of.[2] Even when we discuss an enervated desire to "unplug," we generally imagine a return to a fully electrified life. (The very metaphor "unplug" is impermanent: it evokes an image of being able to reset by eventually reconnecting.) Barring vacations, nightmares, and "natural disasters,"[3] we maintain the sense that electricity runs seamlessly behind the scenes—that we, as consumers, have full control over how we choose to use it. This idea is not written in the pages of a novel or a poetry chapbook, but it is a fiction nonetheless. It is a story that we tell ourselves so that we don't have to confront our reliance on one another and on systems that we cannot personally control. The history of electricity is thus, in part, a story about vulnerability and the techniques we have found to suppress it.

One of those techniques used by Toomer and by twenty-first-century Americans seems merely rhetorical but actually contours the way we comprehend and represent electricity: metonymy. In fact, as I hinted in previous chapters, metonymy plays a significant role in the way we conceive of *technology* in general. Although we cannot escape our reliance on metonymy, there is much to be gained by recognizing that these words only tell a small part of a much broader story. As George Lakoff and Mark

Johnson argue (1980, 37), "Metonymic concepts (like THE PART FOR THE WHOLE) are part of the ordinary, everyday way we think and act as well as talk." They illustrate the function of metonymy with the example of the face as a stand-in for the person. If you see a picture of a person's face, Lakoff and Johnson note, "You will consider yourself to have seen a picture of him"; a picture of a body without a face would not suffice. Similarly, as I (and others) have argued, we consider ourselves to have seen technology when we see a material aspect of a larger, more diffusive sociotechnical system. That small part stands in for a much larger whole when, in turn, we see technology and consider ourselves to have seen progress.

More specifically, when the word *electricity* evokes the components that users see and control—the light you turned on, the power whirring through your devices—it functions as a metonym. Even if the word calls up transformers or power lines, it remains unlikely to elicit an image of the people and artifacts whose actions created, installed, and maintained those apparatuses. Metonymy thus plays an active role in promoting what I have called *the technological fallacy*. It erases the fact that we regularly rely on people and artifacts we rarely (if ever) see; it accentuates only the parts of the bewildering whole that foster a sense of individual control. In short, it encourages reductionism.

Our linguistic habits have evolved in feedback with our cultural and socioeconomic emphasis on individualism. These words emerge from prevailing ideas about life and technology, and they reinforce those ideas in turn. The entire sociotechnical apparatus that we colloquially and metonymically refer to as *electricity* allows us to write for ourselves the fantasy of an orderly world—a world that responds to the buttons we press. Although we use the concept of the "electrical slave" sparingly now, the fantasy of our mastery remains.

BLENDING LITERARY AND TECHNOLOGICAL HISTORY

These metonymies—and the illusions that were built upon them—reflected and reinforced the simplification of electricity as a concept. While Mumford and Ellison were championing a nuanced and humanistic understanding of this energy and its applications, more powerful actors were also at work shaping the social meanings that eventually became dominant. As Hughes (1983) and Misa (2004) have argued, electrical industry executives and investors turned their attention from innovation to system stabilization during and after the 1920s. Financiers began revising the story of electricity. By promoting the idea of "electrical consciousness" that I discussed in chapter 5, these figures re-inscribed the tension between individuals and systems as an issue of consumer choice. As Nye has shown (1997, 384), "Americans used the flexibility of electrical power to atomize society rather than to integrate it. Electricity permitted them to intensify individualism, as they rejected centralized communal services in favor of personal control over less efficient but autonomous appliances."

In light of these developments, the conceptions of electricity that I discussed in the foregoing chapters might seem distant or even naive from our vantage point. Knowing what we now know about electrification and about the rise of neoliberalism in the United States, it can be difficult to fathom why great thinkers such as Gilman and Steinmetz ever believed that electrical interconnection would help Americans grow beyond an individualistic worldview, or why Ellison and Mumford imagined that a neotechnic age might (with some goading) be more humanistic and less utilitarian than the industrial era. Even if we relate to their hopes and fears, we can find it challenging to imagine a different material culture. Would a twenty-first-century American layperson read London's electric pastoral as a truly viable alternative? Or would she be more likely to think that we use power from distant stations because this configuration is the most rational option?

If these ideas about the alternative meanings or configurations of electrical power seem outlandish in hindsight, that is because readers in developed countries have been conditioned to forget about certain aspects of electrification—especially including the human and machine labor that generates and transmits this power. Electricity has not lost its importance in contemporary American culture. It has simply become an infrastructure as Edwards defines the term: it is one of the many systems which is "largely responsible for ... the feeling that things work, and will go on working, without the need for thought or action on the part of users beyond paying the monthly bills" (Edwards 2003, 188). Indeed, this quick mention of bill paying characterizes Americans' current relationship with electricity well: we tend to consider it a purchasable commodity, rather than a flexible system that can be shaped by public debate or by aggregate expert and inexpert decisions. Current discussions about sustainable power generation open up a part of this black box, but not the whole.

These blind spots are exactly why literature should play an important part in broader discussions of technology and society, and why technology scholarship should play a more prominent role in the study of literature. Technology scholarship can help literary critics understand historical contexts and contingencies. In the process, it can help literary scholars ask new questions about authors or texts that have seemed deceptively stable (including ideas of Mark Twain's technological anxiety or Jack London's anti-modernism). At the same time, literary narratives can estrange the language we use to understand technologies, drawing attention to metaphors and metonymies that are more difficult to recognize in technical or colloquial speech. They can also reveal dreamed-of and unrealized possibilities. As Misa explains (2004, 264), the excavation of these alternative perspectives matter: "we lack a full picture of the technological alternatives that once existed as well as knowledge and understanding of the decision-making processes that winnowed them down. We see only the results and assume, understandably but in error, that there was no other path to the present. Yet it is a truism that the victors write the history, in technology as in war, and the technological 'paths not taken' are often suppressed or ignored."

By studying literary and technological history together, we can recover some of this variety and interpretive flexibility that has since become invisible. While ethnographers can analyze laboratories and historians can parse through archives, laypeople have accessed (and continue to access) alternative possibilities most readily in fiction—filmic and televisual, as well as novelistic. I focused on work by prominent writers that fell along a spectrum in the realist tradition because I found these texts to be particularly ambiguous and evocative,[4] but much remains to be discussed. Readers and writers continue to engage with alternative sociotechnical futures across a wide range of genres, from Steampunk through Afro-futurist literature. Examining these intersections, as I have shown, can enrich existing understandings of literature and technology alike.

Conclusion

LOCATING LITERATURE IN OUR SOCIOTECHNICAL IMAGINARIES

Of course, the flexibility that we find in literature is not always the kind that SCOT scholars have sought. Among the professional authors I examined in previous chapters, only Mark Twain described electrical system designs and only Jack London suggested that alternative sociotechnical formations were available before long-distance power transmissions stabilized. The others mainly tinkered with ideas. Gilman opened the stabilized concept of the home and asked her readers to re-imagine this unit as a part of a larger system. This re-interpretation allowed her to promote a kitchenless design and to advocate for men and women to share the governance of the systems that extended the presumably self-contained "domestic sphere." Atherton and Dreiser interpolated the electric chair into a larger system that included journalists and jail time, along with circuitry and power generators. Ellison's and Mumford's contributions were the most abstract; they strove to decouple modernization from utilitarianism, to imagine how electrification might look in a society more open to humanism and romanticism.

Whether they described realistic electrical systems or focused mainly on human interactions, all these writers strove to inspire and were simultaneously inspired by what Sheila Jasanoff (2015, 4) calls "sociotechnical imaginaries," or, "collectively held, institutionally stabilized, and publicly performed visions of desirable futures, animated by shared understandings of forms of social life and social order attainable through, and supportive of, advances in science and technology." As Jasanoff and Kim (2009) theorize the concept, these imaginaries stimulate the development of policies and technologies that might help to realize those aspirations. These imaginaries also underwrite fictions. Authors engage sociotechnical imaginaries when they attempt to shape their readers' ideas about America's electrified future. At the same time, writers inherit their sense of the good life from the sociotechnical imaginaries that pervade and give meaning to the cultures and systems they imaginatively assess. The literary record can thus allow us to examine how a prevailing imaginary took shape—it can, for example, help us understand the American fascination with technologies that enhance atomization and to consider the alternatives that might have been and might yet be.

Nevertheless, as it has been formulated previously, the concept of the sociotechnical imaginary leaves little room for the blended study of literary and technological history. For example, Jasanoff cites science fiction to demonstrate that a culture's anticipatory discourses can precede technological innovation, but she looks to other artifacts to find more tangible traces of sociotechnical imaginaries. Perhaps literature doesn't promote a singular, resilient vision for America's future that comports with Jasanoff's emphasis on durability. Indeed, the most compelling fictions tend to problematize rather than stabilize prevailing ideas about institutions, technology, nation, and selfhood.

The scholar of science studies Stephen Hilgartner coined a phrase for the smaller collectives that promote distinctive visions for the future: "the sociotechnical vanguard." But this concept—though important—doesn't consider the role that fictional narratives can play in inspiring, codifying, creating, or resisting sociotechnical imaginaries. Hilgartner (2015, 34) focuses on the actors who have the technical expertise to "realize particular sociotechnical visions of the future that have yet to be accepted by wider collectives." Literary writers do not play a part in his conceptualization of sociotechnical imaginaries, because they typically don't have the skills to transform their visions into a present material reality. They may not even want to. But the aims of individual authors are beside the point. Like electricity itself, works of literature elicit a wide range of responses that exceed the intentions of the people who created them. Just as users can reshape technologies,[5] readers can reshape the meanings and uses of texts. Narratives that once seemed merely entertaining or escapist might, from another perspective, encourage readers to ask new questions about their imaginaries and their material realities. Without even meaning to, stories help us refine and re-imagine our understanding of the past and our hopes for the future.

The impression that literature leaves on our collective sociotechnical imaginaries may be indefinable, but it is not negligible. I propose that the ambivalences that I discussed in previous chapters reveal the diffraction patterns between emergent and existing visions. Twain explored the tensions that arise when dreams about enhancing individual control with electricity grate against fantasies about improving democracy with the same energy; Ellison examined how that fantasy was complicated by the racialized imaginary which associated electricity with white power. These

Conclusion

tensions caused conflict within each of the narratives I examined. Still, I contend that ambiguousness was not a problem that demanded synthesis or rectification. Rather, this indeterminacy was productive: it influenced how my historical actors envisioned their nation's electrified future.

Like twenty-first-century philosophers who see themselves in and define themselves against artificial intelligence or information technologies, writers and thinkers of the nineteenth and twentieth centuries perceived electricity as a Fun House mirror that reflected different facets of human life. Electricity and humanity shared a similar place in prevailing cosmologies of the day. Both humans and electricity seemed to be of Nature, yet exceptionally set apart from it. Poems such as "Her Lips Are Copper Wire" encouraged this correlation. More concretely, as I discussed in chapter 3, the discovery of bioelectricity and the uses of electricity in medicine encouraged Americans to re-imagine their bodies as power systems and, in turn, to understand power systems as nerves. The traffic between the concepts *body* and *electric* was a cause for celebration, inquiry, and alarm.

When you see yourself in a technology in the way that *fin de siècle* Americans saw themselves in electricity, engineering projects raise some philosophical questions. Is it more comforting to feel that you have control over an orderly universe, that you can tame even the most dangerous natural energies? Or is it more reassuring to believe that some things are exceptional, in the way that vitalists and spiritualists construed humans to be? As Lewis Mumford keenly recognized, *fin de siècle* thinkers were pulled between these two poles. Romantics spurned rationalization, but their model of the universe meant accepting a lack of control that many Americans were loath to tolerate. Utilitarians aspired to quantify, tame, and thoroughly comprehend electricity. Doing so allowed them to combat superstition, fear, and vulnerability. But their model also came with an existential concern: the conceit that we can understand everything comes with the fear that we will forever lose the thrill of discovery.[6] Many literary and technical inventions were born from the attempts to settle this debate; together they suggest that the value of these questions lay not in their determination, but in the process of asking them.

NOTES

INTRODUCTION

1. On the slow development of electrical power, see Nye 1992, 47; Schivelbush 1988, 9–60. More generally, the advent of a new technology doesn't imply its diffusion into daily life. See Edgerton 2006, 4–7.

2. Steam had been the subject of related mechanical fancies, but its association with electrical power diminished in the late nineteenth century as the latter became associated with action at a distance. Although the connotations of these power sources diverged, they were used side by side for heat and power for generations. For a historical account of this relationship, see Steele 1895, passim.

3. See Tresch 2012, xii, 3.

4. In addition to the historically supernatural connotations of electricity, electricity was used to illuminate crosses on churches that attempted to attract followers. See Nye 1992, 51–52. Fictional accounts of electrical spectacles, including notably the World's Columbian Exposition in 1893, also inspired stories that compared electrical displays to heavenly lights. See, for example, Burnett 1895, 78–115.

5. Leo Marx, one of the most influential scholars to discuss the interplay between literature and technology, emphasizes the importance of the pastoral mode in both *The Machine in the Garden* (1964) and *The Pilot and the Passenger* (1988).

6. Clifford Siskin (2016, 20) traces the fascination with systems back to 1610, when "system made a new kind of knowledge while playing an instrumental role in forming our own epistemological practices." His methodologically innovative and deeply insightful approach seeks to understand system as a genre. Here I am interested in a different genre of systems. I study the style of "systems thinking" that examines how perturbations within a system affect a whole, rather than tracing how systems have been used to organize (and thus constrain) knowledge in general.

7. The *Oxford English Dictionary* attributes the first use of the term *system building* to the development of a theory of electricity and magnetism in 1754. At the turn of the twentieth century, writers used the term *system* generically to describe any complex interacting elements that formed a larger whole: they used the term indiscriminately to describe social interactions, mechanical interactions, rhetorical interactions, or interplay among these and other categories.

8. According to the *Oxford English Dictionary*, the word *interrelate* was coined to describe the function of cells working together in a biological organism, *interconnect* was coined to describe the component parts of a curriculum. These words were not initially used to describe the workings of electrical systems, but electrical systems quickly became an analogy for other forms of interconnection and interrelation. The word *network* has a different history. In present-day usage, it conjures an image of abstractly mediated interrelationships. When first applied to electrical systems, however, it was often hyphenated (*net-work*), and it implied a more specific analogy to nets, with all of the affiliated connotations: electric power systems would ensnare regions in an expanding tangle of wires, while some "things" (people, areas, objects) would escape capture. The term *network* registered the fact that some people were disconnected from power and communications lines that connected others, and it called up images of physical structures and wiring that were displaced by discourses that described electricity itself as "invisible."

9. As Aaron Sachs argues (2007, 12), "the word 'ecology' was coined, and the science of ecology invented, by Ernst Haeckel, a German Darwinian, in the 1860s. Yet Humboldt had clearly started thinking ecologically in 1799," claiming that "The discovery of an unknown genus seemed to me far less interesting than an observation on … the eternal ties which link the phenomena of life, and those of inanimate nature." The "chains of connection" that Sachs explores was an important antecedent to the type of "network" or "systems" thinking that this book explores. Whereas I discuss the formation of specific disciplines here, Siskin (2016, 57–63) demonstrates how the very organization of knowledge into disciplines followed from the Enlightenment-era understanding of systems.

10. The formal discipline of sociology was established in 1895. Arguably, the impetus to study social systems emerged from the eighteenth-century construction of the individual as a separate entity from society. On the construction of the individual and its relationship to social theory, see Mumford 1968. On electrovitalism, see Clarke 2001, 151, 190–192; Tresch 2012, xi–xvii.

11. Bertrand Russell discovered his famous set-theoretical paradox—that the set of all sets cannot contain itself and therefore cannot be considered a set— in 1901. That finding disrupted his quest to establish an axiomatic foundation for mathematics. He published *Principles of Mathematics* in 1903, but his discovery created a break between classical and modern mathematics. In 1931, Kurt Gödel famously elaborated on Russell's paradox, and on similar contradictions

discovered by other mathematicians, with his incompleteness theorem. Both discoveries hinted at underlying complexities in the formulation of mathematical knowledge.

12. My questions concerning classification in literary criticism have been inspired by Nancy Glazener's discussion of literary realism and naturalism in *Reading For Realism* (1997) and by Mary Poovey's article "The Model System of Contemporary Literary Criticism" (2001).

13. By associating Whitman with London (author of *The Call of the Wild*), Gilman highlights a congruence between two literary modes that later critics would see as disparate: Whitman's Transcendentalism and London's naturalism both emphasize the theme of man confronting nature. To Gilman, this thematic similarity seemed more salient than the generic differences that later scholars would emphasize in their accounts of American literary history.

14. In view of the fact that the term *infrastructure* was coined in 1927, Gilman probably didn't have that specific concept in mind as she described the link between individuals and municipal systems.

15. For strong readings of Adams's anti-modernism, see Tichi 1987, 137–168; Nye 2003, 261–282.

16. For more on the place of women in Adams's *Education* outside of his largely symbolic depictions in "The Dynamo and the Virgin," see Fleissner 2004, 2–5.

17. For alternative interpretations of the hybrid image of dynamo and virgin, see Tichi 1987, 156; Marx 1964, 347.

18. On the "venus electrificata" experiments, see Delbourgo 2006, 30–31; Rigal 2003, 23–46.

19. Adams did attempt to formalize his epiphany about force in his 1909 essay "The Rule of Phase as Applied to History," but that piece is obtuse and has often been overlooked by scholars. Nye, however, offers an illuminating discussion of this work in the context of Adams's broader catalog on pp. 261–267 of *America as Second Creation* (2003). Mumford discusses William James's interesting response to Adams's essay on pp. 111–113 of *The Golden Day* (1968).

20. Several scholars have problematized the timeline by which modernism displaces the earlier movement of literary realism. See, for example, Lukács 1963; Fore 2012.

21. On the importance of user decisions at the moment of purchase, see Cowan 2012, 253–272.

22. On literature as evocative or formative, see Mumford 1968, 65.

23. On the conventional construction of a case study in sociology, see Ragin and Becker 1992, 3–15; on its construction in intellectual history, see Daston and Galison 2010, 26–27.

24. On the evolution of the term *technology* in American culture, see Kline 1995, 194–221; Marx 1997, 965–988; Schatzberg 2006, 486–512.

CHAPTER 1

1. As Joe B. Fulton notes (1997, 91), the depth of the novel's satire has been a subject of debate: "Although Twain seemed to intend the book as a simple critique of the Medieval period, Hank's violent use of nineteenth-century technology reflects, some have argued, a criticism of his own age as well."

2. For example, Thomas Inge's "Introduction" to the 2008 Oxford World's Classics Edition invokes this mythology in its second paragraph: "Always interested in new devices and mechanical inventions, Mark Twain was particularly taken with a machine that, had it worked, would have revolutionized the printing industry. ... By the time of *A Connecticut Yankee* in 1889, he had for the previous forty-four consecutive months been sending Paige $3,000 a month, and was feeling the strain badly. ... While the exact influence of these unhappy circumstances on Mark Twain's creative work in progress cannot be documented, it seems likely that it would contribute little to a firm faith in American technological progress and economic practices, subjects under scrutiny in *A Connecticut Yankee*" (Inge 2008, vii–viii). The earliest introduction to mention *technology* that I have found is Hamlin Hill's, in the 1963 edition, published by the Chandler Publishing Company. Hill introduces the term conscientiously. He first describes the novel as demonstrating "an industrial revolution based on *economic* progress" (Hill 1963, xvii, emphasis added). Later in that edition, Hill carefully interweaves the notion we would call "technological progress" today: "These notations suggest that Twain himself gave his assenting vote to a technological society which was somehow to precipitate the millennium" (xvii). Even here, "somehow" suggests that Hill is skeptical about how technologies would cause change.

3. See Smith 1964a, passim; Cox 1960, 89–102; Cummings 1960, 17–24. For more recent examples of scholarship that touches on this theme, see Camfield 1994, 81; Seltzer 1992, 11; Fulton 1997, 89–91; Shanley and Stillman 1982, 267–289; Dobski and Kleinerman 2007, 599–624.

4. During the mid nineteenth century, *technology* was used to signify the *study* of practical arts. The word only became generically associated with all of the mechanic arts and with large-scale socioeconomic systems after World War I, largely through the writings of Thorstein Veblen. Eventually it became associated with the material results of narrowly defined techniques. See Marx 1997, 966; Schatzberg 2006, 487. Interestingly, the concept *literature* followed a similar progression: first it was defined as a practice and later as a material object. On this process of reification, see Williams 1977, 47–48.

5. This anonymous writer imagines that peace might only be ensured by Mutually Assured Destruction. While this idea has been widely satirized in

twentieth-century culture, Mark Twain seems to support it in his personal correspondence with Nikola Tesla, a photocopy of which can be found at the National Museum of American History in the *Swezey Papers* (Mark Twain folder, box 17).

6. Many of the scholars who introduced the word *technology* to Twain scholarship also contributed to the formation of American Studies programs. Perhaps they were drawn to this concept because *technology* was associated with the spread of Western capitalism and with fantasies of American exceptionalism. As such the term might seem to tie the novel's disparate themes (such as colonialism, political economy, humanitarianism, fantasy) together. Notably, this was also the decade in which Jacques Ellul published *The Technological Society* (1964), a bleak account of technological tyranny.

7. Here I draw on Bruno Latour's 1996 article "Do Scientific Objects Have a History?" In that article, Latour challenges the notion that Louis Pasteur's "discovery" of lactic acid revealed a scientific object that had *always* existed as such. Although *technology* is not a "discovered" scientific object as such, it has been projected into the past in a similar manner.

8. The intentional and affective fallacies were outlined by a group of literature scholars known as the New Critics, who derided the tendency of literary scholars to project ideas about the author's intention or hypothetical emotional reaction of readers. These "fallacies" have been criticized and elaborated upon. Jane Thrailkill (2007, 1–17) invokes present-day research in the field of affect studies to reclaim a place for embodied, emotional responses in literary criticism. Despite the relevance of such recent work, Wimsatt and Beardsley's work has cautioned generations of critics to reconsider the types of claims that can be made about a literary artifact; I hope the "technological fallacy" serves a similar function, encouraging reflective reconsideration when it comes to our overuse of the word *technology*.

9. The evolving connotations of *technology* were politically charged. Among the scholars who discuss technology from a gender studies perspective, Oldenziel (1999, 44–46) specifically addresses Veblen's role in introducing *technology* with these limited masculine connotations into American discourse.

10. I specify "natural sublime" to differentiate this metaphor from later scenes that David E. Nye has situated in the tradition of the "technological sublime." Although I might modify the adjective *technological*, the tradition of exploring how electricity and other systems could elicit sublime responses is relevant for understanding the spectacles Hank stages throughout this novel. The term *technological sublime* has a long history, dating back to Perry Miller's posthumous book *The Life of the Mind of America* (first published in 1965). My use of this term draws on Nye's rigorous exploration of the concept.

11. Page numbers refer to the 1996 Oxford University Press edition.

12. On the late addition of the "Beginnings of Civilization" chapter, see Smith 1964a, 55–56.

13. For more on historical responses to blackouts, see Nye 2010 and Edwards 2003.

14. I use the terms *romance* and *realism* much as Twain's contemporaries did: *romance* to signify a popular novel, stocked with generic heroes and villains, that probably emphasizes sentimentality and has a happy ending, *realism* to denote attention to mimetic detail. Recent work on American literary realism productively blurs these boundaries. For example, Thrailkill discusses how affects associated with sensationalism could be deployed by literary realists to describe bodies or to elicit bodily responses from readers. I engage with this body of genre theory in my fourth chapter. For the purposes of this chapter, "realism" simply connotes descriptive, mimetic, believable prose. See Thrailkill 2007, 18–53.

15. As Bruce Michelson argues (1991, 615), Mark Twain's "conception of 'realism' seems no more precise than the intention to convey truthfully the outward and inward experience of more or less ordinary folk." When I claim that Twain deviates from or adheres to a premise of realism, I refer to this generic conception rather than a specific literary definition.

16. In March of 1884, several months before recording this journal entry, Twain hired Edward W. Kemble to illustrate *Huckleberry Finn* after enjoying the cartoonist's two-page comic about practical jokes involving hidden electric wires. "Some Uses for Electricity" pictures pranksters shocking whining cats, potential robbers, and even preachers. See Michelson 2006, 266n8.

17. Edgar Allan Poe's 1850 "Some Words with a Mummy" is another story in the American literary tradition that pokes fun at the idea, extrapolated from Luigi Galvani's experiments with a frog, that electricity might awaken dead flesh. Like *A Connecticut Yankee*, this story lambastes American conceptualizations of progress, suggesting that the only new development that distinguishes America from ancient Egypt is the advent of commercialized patent products. For more on this literary tradition that Twain invokes, see Halliday 2007, 7.

18. Carolyn Marvin (1988, 17–32) discusses the use of jokes in nineteenth-century trade literature to distinguish insiders from outsiders in this manner. Mark Twain played with different dimensions of this joke throughout his career. According to the stage he and William Dean Howells set for their play *Colonel Sellers as a Scientist* (with the help of Thomas Alva Edison, who created the electric stage props), electricity is a symbol of science that can easily be misapplied to quackery. Twain also pokes fun at the bumpkin who doesn't understand basic principles of electric conductors and insulators in his short story "Mrs. McWilliams and the Lightning" (Twain 1899, 299–307).

19. I use the term *modern science* colloquially to signify the knowledge and practices that the Yankee uses to dupe his sixth-century spectators, as when he

explains: "Somehow, every time the magic of fol-de-rol tried conclusions with the magic of science, the magic of fol-de-rol got left." Twain tends to use the word *modern* to signify changes in the English language and in American values. His uses of the terms *modern standards* and *modern improvements* in the novel suggest that he measures the past against the abstract metric of modernity.

20. These economic metaphors pervade the novel alongside the electrical metaphors on which I focus. The link Twain draws between financial success and inventions speaks in interesting ways to what Alan Trachtenberg called "the incorporation of America." I will address a link between economics and the emergent concept *technology* in chapter 2. For more on Twain's economic imagery, see Smith 1964b, 77–112.

21. For an interesting take on this question from the perspective of a historian of technology, see Kasson 1999, 211–213.

22. Film adaptations of the novel, such as the 1949 Paramount Pictures version (directed by Tay Garnett) refer to Ben Franklin by name.

23. On the imprint left by Franklin's lightning rods, see Riskin 1999, 61–99.

24. See also Turkle 2011.

25. The title "The Boss" relates to an important economic thread in this novel that I discuss in the conclusion to this chapter. On this moniker and the broader capitalist themes in the novel, see Banta 1991, 487–520.

26. See Plotnick 2012, 825.

27. On didactic electrical displays, see Delbourgo 2006, 100–127.

28. The "disaster shows" Rabinovitz discusses postdate *A Connecticut Yankee*, but the Yankee's perception of destruction as entertainment resonates with her historical account of these shows in *fin de siècle* amusement parks.

29. Twain famously relished the feeling of physically affecting his audience in a similar manner. See Ryan 2009 192–213. See also the analogy Twain draws between authorship and electric circuits (1940, 174): "There is an invisible wire leading from every auditor's soul straight to a battery hidden away somewhere in that preacher's head."

30. Nye (1992, 144; 1994, 278) invokes this very scene from *Connecticut Yankee* to demonstrate Twain's understanding of the "technological sublime."

31. Hank mentions his match factory in chapter 8. On electricity as magic, Jane Gardiner (1987) discusses how Twain uses electricity to position science as "real magic." Gardiner's article analyzes some interesting uses of electricity in the novel, but it overly mystifies electricity. Gardiner's claim that it "could not be readily understood" overlooks Twain's descriptions of basic circuitry that appear in tandem with his spectacular electric displays.

32. On how American authors, including Twain, integrated this nerve science into their understanding of their own craft, see Knoper 2002, 715–745.

33. See, for example, Youngberg 2005, 315–332. On the related theme of imperialism, see Wandler 2010, 33–52; Rowe 1995, 175–192; Shulman 1987, 144–170; Sewell 1994, 140–153. On the relationship between this violent subtext and Hank's depictions of his sixth-century spectators as "an ignorant race" or as white Indians, see Driscoll 2004, 7–23.

34. See also Edison 1885, 185. General Electric continued to develop the notion of electrical slaves throughout the early twentieth century. I touch briefly on Twain's use of this language in my article "Hank Morgan's Power Play" (2010, 61–75). Note that, despite their distinct connotations, the terms *servant* and *slave* have been used interchangeably in the literature of slave owners and in discourses about technological labor. Like Twain and other technological utopians, Arthur Bird made similar claims in his 1899 novel *Looking Forward*; he classified electricity as the "slave of the twentieth century," projecting its usefulness both into the future and the past. On the concept of "wired help," see Lee 1989. Eglash 2007 is useful on the use of "master" and "slave" as component parts in clocks and other devices, but not on the broader uses of the slave metaphor in the history of technology. Carol Pursell briefly mentions the electrical slave on page 266 of her 1995 book *The Machine in America*.

35. Rosalind Williams (1990a, 100–101) also discusses this theme, arguing that nineteenth-century narratives of electricity built on "a particular social fantasy: the possibility of a social transformation through a crucial discovery."

36. See, for example, "Don't be a slave all your life" in the Reddy Kilowatt Collection of the National Museum of American History. Nye discusses advertisements of undercompensated women more generally in *Electrifying America* (1992, 270).

37. The Yankee demonstrates both the hope and the failure of a process we now call "technology transfer." For examples of this practice that were contemporaneous with Twain, see Hughes 1983, 47–78.

38. As Knoper suggests (1995, 124), this convergence of emotional and electrical energy echoes Twain's biographical understanding of "vital force" as a "finer form of electricity."

39. I use *political* in Langdon Winner's sense of the word. See Winner 1986, 19–39.

40. Mark Seltzer (1992, 11) elaborates upon this idea, arguing that the mass electrocution at the end of the novel "posits an identity between signal and act and an identity between communication and execution—'execution' in its several senses."

41. On this notion of progress, see also Kasson 1999, 183–189.

42. See Pfitzer 1994, 42–58; Slotkin 1994, 121.

43. On the harmfulness of romantic literature, see Howells 1887. Nancy Glazener also discusses this conceit in her study of romantic revivalism (1997, 147–201).

44. This fantasy was common during the era in which the novel was published. In fact, Twain's historical model for the electrical wizard—Thomas A. Edison, "the wizard of Menlo Park"—helped Americans reconcile their nostalgia for an age of American individualism and their ambivalence about the modern age of systems. See Trachtenberg 1982, 66.

45. Although Twain doesn't mention the word *bacteria*, the novel was published just as the earliest notions of contagion through bacteria were being formulated. On the development of this new understanding of communicable disease, see Cunningham and Williams 1992.

46. Depending on the reader's familiarity with electrical science, this moment might not be the first one in which Hank appears to be figuratively electrical. Like vernacular descriptions of Ohm's Law, Hank consistently produces his most potent "effects" (often in the form of shocks or sparks) when faced with strong resistance, and he can no longer produce such effects when he is disconnected from his social and technological networks. According to Ohm's Law, $V = IR$, where V is the potential difference or voltage, I is the current through the resistance in amperes, and R is the resistance measured in ohms. In other words, electricity only flows across a circuit and produces effects when forced through a resistant material. Although this algorithm may seem esoteric today, this phenomenon was commonly explained in the popular press at the time. See, for example, "Electrical Resistance," *New York Times*, July 16, 1882; "Nature of Electricity," *The Youth's Companion* 61, no. 6 (1888): 70. Clemens's repeated use of the word *effects* recalls such descriptions of electricity.

47. Note that Twain uses *capital* but not *capitalism* in the novel. See Smith 1964b, 7–8, 39.

CHAPTER 2

1. Salient documents about this movement are conveniently collected in Vila and Morris 1997. According to the timeline put forth by Vila and Morris, the "debate begins anew" in 1977. Although capital punishment had dissenters in intervening years, the movement for abolition lost momentum before the Civil War and didn't return in a highly visible, consolidated form for more than 100 years.

2. Brandon (1999, 12) cites the Smith accident as the "instigating incident" that led to the invention of electric execution. As I will show below, the idea had already been proposed, first by Benjamin Franklin and later by *Scientific American*.

3. See Thomas de la Peña 2003, 113.

4. On the "Battle of the Currents," see Brandon 1999; Essig 2003; Juma 2016. Literary and cultural historians also drew attention to this corporate feud—see, for example, Armstrong 1998, 32.

5. On Edison's competitive reasons for supporting electric execution with AC power, see also Misa 2004, 146.

6. By placing the methods of execution into the hands of experts, removing capital punishment from the public sphere, proponents of capital punishment were able to reframe the practice as scientific rather than atavistic. This understanding of execution persists today in the 32 states that continue practicing the death penalty—including Tennessee, which announced in 2014 that it would return to using the electric chair when sodium thiopental cannot be obtained for lethal injection.

7. See, for example, *Electric Power* 1 (August 1889), which features on pages 255–256 a poem, reprinted from the *New York World*, that ridicules the notion of electric death: "Will Electricity Kill?" Sir Arthur Conan Doyle (author of the Sherlock Holmes series) satirized the electric chair in an 1892 short story, "The Los Amigos Fiasco," in which a convicted criminal is revitalized by a supercharged electric chair. Other notable figures who poked fun at the notion of electric execution emphasized the possibility of mistaking suspended animation for death, raising questions about whether the electric charge or the subsequent autopsy knife would kill those who experienced a "shock."

8. "The Execution of Kemmler by Electricity," *The Electrical World*, August 16, 1890, 100.

9. For this reason, I do not expand my chronology of this chapter further to include Richard Wright's 1940 novel *Native Son*, which was loosely based on the 1939 execution of Robert Nixon.

10. In 1908, after originally being published by The Bodley Head, Atherton's book was reprinted by Macmillan, a larger publishing company, which kept the book in print for years. *An American Tragedy* has consistently been in print since its publication, and has been adapted for the stage and for film. On the publication history of Dreiser's novel, see Pizer 2011, xxi. Interestingly, Atherton was not the only writer to re-invent the Harris case. Harris was executed on May 8, 1893. On June 17 of that same year, in the Windsor Theater in New York, a performance titled "Life and Death of Carlyle Harris, or, the Road to the Electric Chair" was put on. Atherton's work responds to this problematic relationship between criminal justice and theatrical entertainment. In 1966, Henry Hamilton reinterpreted the Harris trial in a short story titled "The Case of the Six Capsules."

11. On the philological debate over the adoption of the term *electrocution*, see Charles Heinrichs, "Disnerving Instead of Electrocution," *Chicago Daily Tribune*, August 14, 1891; "Electrocution," *The Youth's Companion*, May 18 1899, 25; "Elektramort," *Chicago Daily Tribune*, July 28, 1891; "Catelectrize Versus Electrocute," *Chicago Daily Tribune*, August 3, 1891.

12. Robert K. Merton (1968, 14) warned against whiggish claims of "anticipation." I use the word here to note a "resonance" of the sort Wai Chi Dimock and Ellen Spolsky describe in "A Theory of Resonance" (1999). Since Twain published *A Connecticut Yankee* after the Electric Execution Act was passed but before it was first implemented, it is fair to say that he was aware of rhetoric akin to *Scientific American*'s argument. Twain may not have read the *Scientific American* piece specifically, but the congruence between these works warrants reflection.

13. Martin Heidegger's discussion of the ideal that the physical world be "calculable in advance" (1977, 21) offers important insight here. Much has already been written about the stakes of adopting numerical, utilitarian language to moral issues. On the entanglement of economics and discourses of scientific rationality, see Rescher 1996, 6–8; Mirowski 2004, 72–74. For a history of economics that examines the permeable boundary of engineering and physics, see Porter 1995, 49–72. While I agree with these scholars that narratives reify electric execution by reducing it to a problem of economic optimization, what interests me is what this act of reification might indicate about the emerging ideas that would eventually coalesce into the concept *technology*. As Leo Marx argues (1997, 974), "the amalgamation of science and industry helped to call forth the concept of a new realm of innovation and transformative power—a new entity—called *technology*." Although I am skeptical of the conjuring power of circumstances to "call forth" a new concept, I agree that the entanglement of economic concerns and material culture was an important aspect in the evolution of the concept *technology*. For example, Schatzberg (2006, 497) argues that an 1887 translation of *Das Kapital* was one of the first texts to use the term *technology* in the way we use it today. Thorstein Veblen, widely credited with bringing the broader use of the word into common usage, also used the term *technology* in his 1898 article "Why Is Economics Not an Evolutionary Science?" The complex of ideas that brought economics, scientific techniques, and material artifacts into closer correspondence seems implicated in the evolution of this concept.

14. Interestingly, this satirical piece by Howells was reprinted in trade journals. See, for example, *Western Electrician* 2, no. 4 (1888): 38–40 and *The Electrical World* XI.3 (21 January 1888): 27–28.

15. On the cultural meanings of the electric button, see Plotnick 2012, 815–845; Suchman 2007, passim.

16. Note that the illustration for "The Devil's Acre" reproduced here as figure 2.1 appeals to the sensationalism about which Crane's text raises questions. The dissonance between visual and textual rhetoric might have minimized the impact of this piece of cultural criticism.

17. On the meaningfulness of apparently mundane details, see Barthes 1986, 141–148.

18. Stephen Crane's 1898 story "The Monster" explores the ethical dilemma that arises when Dr. Trescott saves the life of a black servant, Henry Johnson, after the man is severely disfigured and traumatized by a fire in the doctor's own lab. Trescott saves Johnson because Johnson had saved his son from the fire, but this combination of mercy and medical magic leads to the ostracization of the Trescotts. Depicting a scene of "cultural lag"—when scientific advancements seem to outpace morality—the short story details the myriad problems townspeople in a small, upstate New York town try to re-interpret their values on an increasingly modernized landscape. Jonathan Tadashi Naito explores this theme in his 2006 article "Cruel and Unusual Light."

19. Crane's depictions resonate with the once-popular but now-discredited theory of "cultural lag" that William F. Ogburn proposed in 1922. Although technology studies scholars have argued persuasively that "technology" is not an autonomous entity that can outpace human morality in this manner, the impression of cultural lag has been no less influential. Much in the way that Latour's *Inquiry* describes the cultural relevance of the concept of modernity regardless of the fact that, in Latour's words, "we have never been modern," the fear that American culture was unprepared for the invention of the electric chair was influential, even if it cannot describe fully the existing cultural forces that contributed to the construction of the device. This sense of fear lends to the rise of the term *technology*: it indicates a predilection for conceiving of the physical invention as acting autonomously from the social and economic contexts in which the device was formulated, designed, and implemented.

20. Crane describes his visit to the electric chair in "The Devil's Acre." On Atherton's visit to Sing Sing, see Leider 1993, 147. On Crane's philosophy of embodied experience, see Stallman 1952, xv–xxi.

21. As Dwight Conquergood argues (2002, 360–361), "Officials are anxious to control the performance because condemned prisoners, although acutely vulnerable, are not without agency. They can fight back and force the guards to drag them kicking and screaming to their death."

22. Brandon 1999 includes an entire chapter detailing Kemmler's trial and execution.

23. This interpretation was further promulgated by the official report that analyzed Kemmler's body after his execution. See MacDonald 1890; also see "The Kemmler Execution," *Chicago Daily Tribune*, August 8, 1890.

24. Less sensational reports than the one published in the *Police Gazette* discussed the desire to optimize electric execution more overtly. For descriptions of the electric chair that emphasize the scientific process, see "Kemmler's Death Chamber: How the Electric Current Is to Be Transmitted to His Body," *New York Times*, April 26 1890; also see "Capital Punishment," *New York Times*, December 17, 1887.

25. On the expectation that the press would lie, see Roggenkamp 2005, 13–16.

26. Interestingly, Davis was the electrician who executed both Carlyle Harris and Chester Gillette, the men upon whom Atherton and Dreiser based their novels. For more on Davis's long career as the electric executioner at Sing Sing, see Brandon 1999, 208. The journal *The Electrical World* also featured the perspective of an expert who wanted to learn more about electricity by witnessing this execution (Charles R. Huntley, "The Execution as Seen by an Electrician," *The Electrical World*, August 16, 1890, 100). Huntley lamented having been unable learn the voltage that had been used to kill Kemmler.

27. On coverage of hanging, see Brandon 1999, 25–46.

28. The doctor who witnessed the first electric execution made a similar claim. An article in the August 16, 1890 issue of *Electrical World* article ("The Execution of Kemmler by Electricity") quotes a medical expert, named as Dr. Shrady: "[T]he long lengthened agony of suspense regarding the efficiency of electricity as a means of executing criminals has finally terminated in the legal killing of William Kemmler.... But shall we call it a triumph when the object attained was the killing of a fellow-being?"

29. The original clause in the Electric Execution Act that prevented members of the press from observing and reporting upon electrocutions was repealed on February 4, 1892 by New York Governor Roswell Flower in the Code of Criminal Procedure (Section 507). See Moran 2003, 213.

30. Atherton discusses her critique of the American fascination with criminal plots in her autobiography (1932, 226).

31. Page numbers refer to the 1953 Modern Library edition.

32. See, for example, Dreiser 1925, 351 and 377.

33. Originally printed in *Mystery Magazine* 1, February 1935.

34. This was a recurring theme in the oeuvre of Dreiser, who had crafted a similar depiction of fallen manhood in his 1900 novel *Sister Carrie*. The character Hurstwood naively believes that he can escape his past wrongs, including his decision to steal money from employer. He tries to escape to Canada with Carrie, but news travels faster down telegraph wires than he can travel by train. Once found out, he slowly descends into poverty and hunger. See Dreiser 2006, 243–354.

35. For more on Atherton's representations of journalism, see Lutes 2007, 104.

36. Page numbers refer to the 1908 Macmillan edition.

37. On audiences' changing attitudes toward to the death penalty, see Boudreau 2006, passim; Foucault 1977, 47–66.

38. The first woman to be executed in the electric chair was Martha M. Place, executed at Sing Sing Prison on March 20, 1899.

39. See Crowley 1999, 92.

40. On the relationship between Dreiser and Veblen , see Eby 1998. On Veblen's use of the word *technology*, see Oldenziel 1999, 44–46.

41. Eby (1998, 3) describes Dreiser and Thorstein Veblen as sharing an "institutional approach." Understanding Dreiser's perspective in these terms can help to elucidate why his narrative voice is notoriously abstruse: instead of describing characters, he wants to describe entire, complex, interconnected systems.

42. Albert Camus poetically elaborated upon this fact in his 1946 pamphlet *Neither Victims Nor Executioners*.

43. Dreiser's pessimism might have resulted from the fact that he wrote this novel as technical systems were stabilizing, rather than changing dramatically. On this move toward system stabilization, see Misa 2004, 148–149.

44. Later in life, Gertrude Atherton demonstrated her adherence to electro-vitalist beliefs by undergoing the Steinach treatment—a procedure in which she exposed her ovaries to x rays in order to revitalize her aging body. (Atherton fictionalized her experience of the Steinach treatment in her 1923 novel *Black Oxen*.) But she cultivated an allegiance to theories about electricity and vitalism much earlier, and she builds these theories into Patience's gradual development throughout the novel. Atherton was proud of the intellectual inheritance of her ancestor, the famous electrical experimenter Benjamin Franklin. She used the pseudonym Frank Lin in his honor. She also published under her full name, Gertrude Franklin Horn Atherton, in order to advertise the relation. Beyond her ancestral interest in electricity, she read the works of intellectuals who perpetuated vitalist representations of electricity (among them Ralph Waldo Emerson), and she prided herself on the scientific perversity that others saw in her. See Atherton 1932, 104. Her biography therefore renders her narrative play with electricity as a multivalent symbol more conspicuous.

45. On the invisibility of infrastructures in the developed world, see Edwards 2003, 188–195.

CHAPTER 3

1. Although Dreiser establishes this "tragedy" to criticize the class system, he also reproduces market logic. See Seltzer 1992, 87.

2. Gilman also used the cell metaphor to describe the smallness of the individual human. In one essay, Gilman (1991a) describes "the individual man merely as a cell in the structure." For other examples of "cells" in Gilman's work, see Gilman 2012, 16–17; Gilman 1991b, 302–303.

3. During the 1890s and the 1900s, more than a hundred utopian novels were published in the United States. See Roemer 1976; Pfaelzer 1984.

4. For more on Camp Co-operation, see Kline 1992, 214–217. For a useful discussion on the relationship between General Electric and Steinmetz's socialist politics, see Jordan 1989, 57–82.

5. For more on Trowbridge's body of work, see Lieberman 2016b.

6. Electromagnetism is the only physical theory that concerns attraction and repulsion.

7. I use scare quotes around the word *hard* here to draw attention to the problems with this value-laden adjective. For a brief, lucid example that problematizes this hierarchy of the "hard sciences," see Malina 2009, 184.

8. See, for example, her 1911 novel *The Crux*.

9. On the liminal materialism of electricity, see Gilmore 2008, 5–10.

10. Whitman uses lists to represent this equivalency. Instead of using the neutral, inclusive language of "humanity" or "all humans," he specifies that he speaks for "Man, woman, and slave," arguably retaining the hierarchies or distinctions his poem ostensibly attempts to dismantle.

11. On Gilman's inheritance from Whitman, see Golden 1999, 243–266.

12. "The Rest Cure" is the treatment Gilman depicted in "The Yellow Wall-Paper," which forced women coping with neurasthenia to eat heavy foods without being allowed to leave their beds or to write. This cure was gendered: men who suffered from similar ailments were prescribed time outdoors, symmetrically named "The West Cure." For more on this treatment regimen and Gilman's responses to it, see Tuttle 2000, 103–121.

13. For more on Gilman's electrical and organic metaphors, and specifically on her use of neurological imagery, see Thrailkill 2007, 125–135.

14. Despite Gilman's criticism of Karl Marx, this theory also resonates with the Marxist account of social life. For example, Horkheimer and Adorno (2002, 125) argue that "no individuation was ever really achieved. ... The individual, on whom society was supported, itself bore society's taint; in the individual's apparent freedom he was the product of society's economic and social apparatus." Also see Gilman 1991b, 302–305. In the passages quoted, Gilman calls to my mind Karen Barad's notion of agential realism or agential cuts in *Meeting the Universe Halfway* (2007); both Gilman and Barad suggest that the idea of the detachable human is a fiction enabled by established strategies of perception.

15. On this interview, see Nye 1992, 242. On Edison's boasting, see Bazerman 1999, 159.

16. Hertha Ayrton was elected into the Institution of Electrical Engineers in 1899, two days after she became the first woman to read a paper to the Institution. In 1906, she won a Hughes Medal from the Royal Society of London for her original research in electricity. Ayrton's work was widely publicized because of her gender. See, for example, "A Woman Electrical Engineer," *New York Times*, March 28, 1899.

17. According to Alan Trachtenberg (1982, 66–67), Edison's public image also served to perpetuate notions of romantic individualism. Although actual electrical systems were developed by a team of researchers, technicians, and construction workers, the public image of Edison as the charismatic genius behind these developments allowed Americans to imagine that individual ingenuity could control and guide these systems.

18. Cowan and Wajcman consider why alternative systems of domestic labor failed. In *More Work For Mother*, Cowan theorized that women adopted electricity into the home because they were influenced by social mores involving privacy (1983, 149–150). In *Feminism Confronts Technology*, Wajcman points out that the interests of electrical-system owners played a powerful role "in shaping domestic technology" (1991, 100). I agree with both assessments, but I would add that the industry-sponsored model of electrifying specific tasks and the alternative model of collectivizing the home both served to reduce the middle-class, white household's dependence on lower-class labor by non-white peoples. This shared aim might have rendered the adoption of electricity especially appealing to genteel white consumers.

19. See, for example, Gilman 1904, 376.

20. See Brackett et al. 1891, 42, 142.

21. This story of electrical development was parroted in various magazines, newspapers, and journals. Practitioners of electrical science told and retold the history of energy consumption, repeatedly celebrating the consumption of energy from a distance that alternating-current systems afforded. For example, Nikola Tesla shared this story with an audience of literary and cultural elite in the highbrow periodical *The Century*: "The ultimate results of development in these three directions are: first, the burning of coal by a cold process in a battery; second, the efficient utilization of the energy of the ambient medium; and, third the transmission without wires of electrical energy to any distance" (Tesla 1900, 193).

22. A trade journal titled *Electrical Merchandising* was founded in 1905 to address the various vested interests in the electrical industry and to suggest that all interested parties should garner excitement for "electrical progress" and

should promote the consumption of electrical energy in whatever form it took. A few snippets can help to illustrate the scope of this obscure publication. Earl E. Whitestone's "To Harness the Whole Industry for You" (*Electrical Merchandising* 17, no. 1, 1917, 5) is about when electric companies coordinate with appliance salespeople they sell more things to "Mrs. Housewife." The same issue included an article, titled "Scattered Efforts Mean Waste," that claimed: "In many communities the public is ready to buy appliances. The desire to 'do it electrically' has been instilled by the manufacturer's popular-magazine advertising and the propaganda of the Society for Electrical Development." And "The New Spirit in Merchandising" (17, no. 2, 1917) made this claim: "The retail distributing of electrical goods outside the technical and applied engineering fields is the weakest side of the industry. There is an historical reason for this. The electrical industry has been built backwards. We started as a science. The names of electrical units are the names of scientists. We have specialized in lighting, in railway, in power, in telephony, in government, and in public service. We have only lately begun to realize that we must also specialize in merchandising." Also in issue 2, an article titled "A Ladies' Day in the Electric Shop" suggested that store owners host a "ladies' day" to promote women's consumption of electrical goods: "An Indiana dealer in electrical supplies holds what he calls "Ladies' Day" once a year. On this occasion the store is given over entirely to the fair sex. From stem to stern the interior is artistically decorated, small cups of cocoa and wafers are served all day, and a phonograph furnishes popular music. … The women, of course, closely inspect the products on display, and the dealer says that actual sales on these days invariably more than offset the expense incident to staging the affair." Interestingly, *Electrical Merchandising* saw hired help as a means to boost sales and not a social issue to be replaced by inventions. The same 1917 issue offered this suggestion: "Convert the maid and she will help you sell it to the mistress, and will advertise you through the neighborhood."

23. Steinmetz does recognize in this speech that users might use isolated generators for their own power; he classifies this form of energy generation as outdated.

24. On load management, see Hughes 1983, 216–219.

25. According to Veblen (1912, 98–100, 397), the wastefulness connoted by women's uselessness was one aspect of conspicuous consumption. Tim Armstrong (1998, 58–73) usefully catalogs the turn-of-the-century discourses about waste and efficiency.

26. Gilman makes this claim on pages 38, 169, 201, and 220 of *Human Work* (1904).

27. For more on Gilman's developmentalism, see Ahmad 2009, 49–65. For a discussion of how developmentalist thought inflected late-nineteenth-century American culture more generally, see Lears 1983, 4–31.

28. Karen Barad (2007, 137) introduces the helpful notion of the "cut" to describe how entangled assemblages come to appear as "individually determinate entities."

29. Perhaps this, and not her eugenicism, was the experience that drew Gilman to explore our systemic connections. As Rebecca Solnit says (2013, 129), when you suffer from illness "you cannot ignore that you are biological, mortal, and interdependent."

30. Beard and Rockwell's *Practical Treatise* (1884, 1–2, 88) describes electricity as easier to understand and less mysterious than the human body.

31. This interpretation of the living body as a sort of powerhouse was popular outside of this specialized profession. For more on popular and technical definitions of entropy and electro-vitalism, see Armstrong 1998, 14–17; Clarke 2001, 74; Mirowski 1991, 61–65, 361; Thomas de la Peña 2003, 27–29.

32. Here I use the term *power grid* anachronistically and playfully. The concept of the electric grid didn't appear until 1926, according to the *Oxford English Dictionary*. In Gilman's day, grid meant a component plate of a storage battery. Thomas Markus (2002, 22–23) also notes how Gilman's utopias (like others of her day) were ordered on grid-like plots.

33. For example, Arthur Morgan, Bellamy's biographer, went on to work for the Tennessee Valley Authority in part because he believed that the project of irrigation and electrification reflected Bellamy's vision.

34. Notably, 14 years after Gilman published *Moving the Mountain*, the Electrical Association for Women—located in London—began to lobby for safer electrical regulations in their homes, much as Gilman foresaw in that utopian novel. For a useful account of this movement in England, see Pursell 1995, 47–48. Many of the publications of the Electrical Association for Women are held by the Bakken Library and Museum in Minneapolis.

35. See Stepan 1986, 261–277.

36. See, for example, Gamble 1997, 1773–1778.

37. On the technological utopia and the electrical utopia, see Segal 2005; Lieberman and Kline 2017.

38. The phrase "second industrial revolution" originally referred to the period from the 1870s through the 1920s, with the inventions of electrification, automobiles, typewriter, radio, and so on.

39. The Giant Power initiative ultimately failed because utilities companies saw interconnection as too much of a risk. See Misa 2004, 155.

40. Interestingly, the electrical hearth would become an important image in General Electric's advertising campaign twenty years later. See Nye 1992, 238–286.

41. Gilman draws analogies between electricity and human life, as in her human storage battery image. She also hints that modern science has re-imagined what it means to be human altogether, perhaps hinting that the discovery of bioelectricity has de-familiarized the human body altogether. In the opening to *Human Work,* she claims: "We had, and used, and supposed we knew, our own bodies, through long centuries of living and dying, yet our late-learned physiology was able to show us facts most vitally important which we had never dreamed of" (1904, 5).

42. Many of the utopians that Howard Segal discusses also had experience in technical trades. See Segal 2005, 45–55.

CHAPTER 4

1. On the uses of autobiography in London's fiction, see Auerbach 1996, 20.

2. On the professionalization of electrical engineering, see Terman 2002, 1792–1800. See also Kline 1995, 194–221; Hughes 1983, 141–157.

3. The development of science and engineering curricula at western colleges was an important move by land-grant institutions that wanted to "call to [their] halls the number of students who are, each year, going to eastern colleges for scientific training, we must keep our departments of science abreast of the times" (Cahan and Rudd 2001, 11–12). Misa (2004, 151) also discusses the changes in engineering education that came at the turn of the twentieth century.

4. For example, see the review of *Burning Daylight* in William Morton Payne, "Recent Fictions," *The Dial,* November 16, 1910, 384. Recently *Burning Daylight* and *The Valley of the Moon* have begun to receive more recognition. For example, Jonathan Auerbach (1996, 229) describes *Burning Daylight* as "perhaps the most interesting" among a thematic trilogy of books including the 1908 novel *The Iron Heel* and the 1909 novel *Martin Eden,* and Laurent Dauphin (1995, 194) writes: "When *The Valley of the Moon* was published in 1913, it was an immediate success. [Although] its importance as a work of art is often minimized … Jack London never failed to assert that *The Valley of the Moon* was one of his favorites among the several novels he had written."

5. See Kline and Pinch 1996, 763–795; Pinch and Bijker 2012, 11–44.

6. Literary scholars also examine interpretations that arise from embodied interactions with literature or with material culture. See, for example, Michelson 2006; Thrailkill 2007.

7. Jonathan Auerbach and Christopher Hugh Gair express concern about the stability of post-Reconstruction and post-frontier American national identity at the heart of London's fiction. See Auerbach 1996, 66–69; Gair 1996, 141–157.

8. On this interpretation of *The Valley of the Moon*, see William Morton Payne's review of *Burning Daylight* in the November 16, 1910 issue of *The Dial*.

9. See Pinch and Bijker 2012, 40.

10. See Glazener 1997, 6.

11. London's conclusion need not be read as an anomalous deviation from the naturalistic body of the story. Sara Britton Goodling (2003, 1–22) argues that sentimentalism and literary naturalism share roots in American literary history.

12. NELA was founded at the end of the nineteenth century to influence public policy by leveraging "the support of the leading utilities executives." See Hughes 1983, 206–207.

13. As Nye argues (1985, 136), "As early as 1906 the topic [of privatization] proved so popular that the Brooklyn Public Library prepared an annotated list of works on the subject, numbering more than seventy-five books."

14. George W. Norris was an advocate of electrifying the nation; he supported both the Tennessee Valley Authority Act of 1933 and the Rural Electrification Act, but he disapproved of the practices of private utilities.

15. On the murky definition of "law" and how it was important for engineers' ideology, see Layton 1971, 53–54.

16. Opinions about the influence of western engineers on the development of power systems differ. A particularly interesting example, a review of current literature in the April 1896 installment of *The Literary World*, calls "special attention" to two articles in *Appleton's Popular Science Monthly*, including "an extremely interesting paper by William Baxter, Jr., on 'The Electric Transmission of Water-power.'" The reviewer situates power transmission systems as equally eastern and western by citing power plants on both coasts. See "Review of Current Literature," *The Literary World*, April 1896, 120.

17. Jack London wasn't the only writer to take up this theme. For example, between May and June of 1896 the upper-middle-class periodical *The Century* published "The Harshaw Bride," a novella that depicts how a power plant might fit into a rural community in the young state of Idaho. "The Harshaw Bride" was reprinted in Mary Hallock Foote's 1903 collection of short fiction *A Touch of Sun*. *Overland Monthly*, a journal of western literature and culture, also included articles about that described electrical developments in terms of western iconography. See Edward Berwick, "Farming in the Year 2000," *Overland Monthly and Out West Magazine* 15, no. 90 (1890), 569–573; Alvan D. Brock, "The Supplanting of Steam," *Overland Monthly and Out West Magazine* 14, no. 82 (1889), 396–409; W. A. Tenney, "Evolution of the Northwest," *Overland Monthly and Out West Magazine* 35, no. 208 (1900), 321–332; Vere Withington, "An Electrical Study," *Overland Monthly and Out West Magazine* 20, no. 118 (1892), 417–429.

18. Although it may seem odd that London uses electrical imagery to describe mail delivery, as Greg Downey argues, the postal service was considered part of the telegraph and telephone information "internetwork" during the early twentieth century. See Downey 2001, 213–214. For more on mail imagery in London's work, see Auerbach 1996, 12.

19. In his influential medical treatise *American Nervousness*, George Miller Beard claims that "in regions where the atmosphere is excessively dry, as in the Rocky Mountains, human beings—indeed all animals, become constantly acting lightning-rods, liable at any moment to be made a convenient pathway through which electricity going to or from the earth seeks an equilibrium" (1881, 147). George Wharton James similarly advertised the electrical atmosphere as a salubrious aspect of western tourism. In *The Wonders of the Colorado Desert*, James lists "seductive electric conditions" among the "Desert surprises," contending that "during the cool of the early morning, while the air is like champagne or some electric fluid coursing through his veins and giving to nerves and muscles unwonted sensations of stimulus and exaltation" (1906, 35).

20. See La Vergata 1995, 193–229.

21. For Edison's discussion of Spencer, see Shaw 1878, 490. On Tesla's, see Tesla 1900, 175.

22. Latour and Woolgar (1986, 88) noticed the surprising amount of inscription that pervaded a scientist's "laboratory life," and they argued that practice of writing preceded and enabled the scientific discovery. In order to gain a consensus that a phenomenon had been discovered or formalized, scientists had to enlist allies with convincing textual arguments. See also Lenoir 1998.

23. By the time London published *The Valley of the Moon*, Thorstein Veblen was already bringing *technology* into the American vernacular. This novel and *Burning Daylight* both appeared between the 1899 publication of Veblen's *Theory of the Leisure Class* and the 1920 release of Veblen's *The Engineers and the Price System*. The latter publication attended more closely to this term, as it sought to describe what engineers created. Veblen's answer is that they made technical knowledge. Even in this latter study, the material connotations of the term *technology* seems an afterthought.

24. See "Cleveland Convention Notes," *Electrical World*, October 29, 1892, 281.

25. Although the version published in *The Electrical Journal* was titled "Long Distance Power Transmission" rather than "Americans Have Great Courage," its proximity to an article titled "Politics and Municipal Lighting" more clearly situated this piece as electric industry propaganda.

26. Whereas I use the singular phrase "electrical industry" for rhetorical reasons, McGhie alternately describes the electrical industry as singular ("the electric

mind") and plural ("the various systems of electric generation and alimentation"). See the *Cassier's* version of this article (McGhie 1896, 359). This division is hierarchal in the article. McGhie, in the *Cassier's Magazine* version, suggests that electrical developments for profit are a "less elevated idea" and disparages "the electrical business" for "coming down to a strictly commercial basis."

27. For more on this issue in American letters, see Michaels 1987.

28. On the racial dynamics of the novel, see Gair 1996, 143–147.

29. "Developments in the West," *The Electrical Age*, July 10, 1897, 23.

30. London did write a socialist fantasy five years earlier: "The Dream of Debs," in which the proletariat gained control over telegraph and telephone lines by cutting them and waiting for the bourgeoisie to recognize their reliance on the labor that keeps these systems running. But even that story didn't demonstrate "interpretive flexibility"; it didn't change the design of these systems; it merely changed the ownership.

CHAPTER 5

1. Lewis Mumford and Ralph Ellison were National Book Award winners for nonfiction and of fiction, respectively.

2. For a notable exception, see Wright 2005, 157–171, or Nadel, 88–95.

3. In other words, I do not attend to Mumford's later critiques of technology *The Myth of The Machine* and *Pentagon of Power;* I am considering instead a moment of his relative optimism.

4. Page numbers refer to the 2010 University of Chicago Press edition.

5. On this debate about whether technologies function within or outside of society, see also Misa 2004, xi.

6. See, for example, Mumford 1934, 4–7.

7. In this sense, Mumford can be said to have revived for the twentieth century the idea of "romantic machines" that John Tresch discusses in his 2012 book *The Romantic Machine*.

8. Page numbers refer to the 1995 Vintage International Edition.

9. On the whiteness and masculinity of engineering, see Oldenziel 1999 and Slaton 2010. Ellison's attraction to technology is evident in his biography as in his fiction. Adam Bradley (2010, 32) notes that Ellison was one of the first authors to eschew the typewriter for the computer, incidentally becoming "a literary pioneer for the digital age." Sara Blair (2005, 21–44) catalogs Ellison's "impressive array of cameras, lenses, and other technical apparatuses" as an analogue for his arsenal of narrative strategies. And these were only a few examples:

Ellison consistently undermined the notion that white experts were the only people capable of controlling or inventing technology.

10. See Lieberman 2015, 8–27. Before I elaborate on this terminology, I should note that I recognize that this focus on the human can seem passé in current scholarship; I acknowledge how important it has been for posthuman poststructuralist, and Actor-Network theorists to dethrone the concept of the human from its exceptional place in the Western philosophical tradition. But at this point I also ask readers who are steeped in these traditions to remember that the allure of humanism couldn't be discarded by Ellison, who had to fight for the status of human.

11. Note that this conventional usage of the word *humanist* is slightly different from the definition Ellison offers in "Twentieth-Century Fiction and the Black Mask of Humanity" (1964, 33).

12. Mumford also focuses on the theme of slavery in his 1930 article "The Drama of Machines," in which "slaves" includes the machines that people use and other alienated forms of human labor..

13. As Nye notes in *Electrifying America* (1992), the federal government took advantage of the public's dependence on these signs by replacing a Wrigley's chewing gum sign with advertisements urging people to support the war effort.

14. "Developing an Electrical Consciousness," Gerard Swope Papers, Museum of Innovation and Science, Schenectady, New York, box 27, folder 2012 . For more on this campaign, see Marchand and Smith 1997, 161.

15. According to the *Oxford English Dictionary*, the word *electronic* was first used to describe a device (and not a theory of physics) in 1919.

16. Mumford's treatment of electricity as exceptional followed from his reading of Patrick Geddes, the Scottish scientist and writer who coined the terms *paleotechnic* and *neotechnic*. See Geddes 1915, 129. Mumford also might have been influenced by prevalent debates in American culture regarding whether machines enslaved humans or humans enslaved machines. On this debate, see Chase 1929, which opens with the master/slave metaphor and pursues this theme throughout the text.

17. Mumford added to Geddes's schema the "eotechnic" age, which enabled him to transform the contest between paleotechnics and neotechnics into a more expansive teleology. See Williams 1990b, 43–65.

18. For example, Mumford cites the mechanical stoker—an invention that rendered Jack London's erstwhile position in the power plan obsolete—as a tool that made electricity more readily available while "emancipat[ing] a race of galley slaves" (235). Although the stoker was mechanical, Mumford claims that it helped to unleash neotechnic energies.

19. On the stereotype of the idle black slave, see Fouché 2006, 647.

20. On Ellison's play with radios, see Wright 2003, 178. On Lewis Mumford's early work with radios and electronics, see Hård and Jamison 2005, 104.

21. On Ellison's use of the computer, see Bradley 2010, 32.

22. Ellison was inspired to use electrical imagery, in part, by reading Ernest Hemingway and F. Scott Fitzgerald. On these influences in Ellison's work, see Wright 2006, 153–157.

23. Ellison expresses his familiarity with the conventional "body electric" imagery in *Invisible Man* when a nameless veteran in the Golden Day bar says "I'm a dynamo of energy. I come to charge my batteries" (81).

24. Page numbers refer to the 1980 edition of *Invisible Man*.

25. Fouché (2006, 639–661) discusses how histories of technology that focus singularly upon the oppression of African Americans fail to recognize subversive types of technological creativity that African-American thinkers and tinkerers used as they operated actively (rather than passively) within an oppressive culture.

26. I draw here on Akira Mizuta Lippit's discussion of "visual aurality and aural visuality" in *Atomic Light* (2005, 100).

27. As I argue elsewhere, this use of the social meanings of electricity to influence white donors was in fact an aspect of Tuskegee Institute's History. See Lieberman 2016, 70–90.

28. For readings of Trueblood as a "signifying" raconteur, see Baker 1983, 828–845; Benjamin 2014, 121–148; Ford 1999, 887–904; Johns 2007, 230–264; Marvin 1996, 587–608.

29. On the use of girls and women in electrical advertisements, see Nye 1992, 244; Marvin 1988, 166; Thomas de la Peña 2003, 106–107.

30. See, for example, Barrett 1894, 23.

31. Ellison's technological humanism, and his insinuation that multiple perspectives should be consulted even when those perspectives are undesirable to people in power, accords with Michel Callon's, Pierre Lascoumes's, and Yannick Barthe's idea of "technical democracy": the process of genuinely including multiple perspectives in the construction of knowledge about science and technology. See Callon et al 2009, 9.

32. Although *The Golden Day* is rarely remembered today, it made an impression on the field of American literary studies by offering a blueprint for F. O. Matthiesen's more influential American Renaissance. See Nadel 1991, 94.

33. *Emplotment* is a term used by Hayden White (1973, 5–11) to describe the way that historians work disorganized archival data into an understandable plot.

White demonstrates how historians use Northrop Frye's archetypal forms: Romance, Comedy, Tragedy and Satire. White describes a romantic emplotment as a story of individual success and "a drama of the triumph of good over evil, of virtue over vice, of light over darkness, and of the ultimate transcendence of man over the world in which he was imprisoned by the Fall" (8–9). When Mumford idealizes the Transcendentalists or the possibilities of neotechnics, he appeals to this convention.

34. Nadel suggests that Mumford projected his feelings about World War I onto his account of the Civil War, rendering Ellison's translation less far-fetched. See Nadel 1991, 88.

35. One of the veterans who speaks truth to power through his insane ramblings describes himself as a "casualty."

36. See Ellison 1964, 282–293; Lieberman 2015, 4–5.

37. He doesn't recognizes the significance of this turning point until p. 576 in the novel.

38. See Benjamin 2014, 123–124; Dixon 1980, 98–104; Tate 1987, 163–172.

39. Other scholars have discussed the violence of this scene at length but have not interrogated the resultant sense of equality with the white doctors who subjugate him. See Wilcox 2007, 987–1009; Curtin 2003, 52–55.

40. The telegraphic imagery is more apparent in the alternate manuscript version, as Invisible Man thinks about asking his doctors "What hath God wrought?"—the first phrase ever transmitted by electromagnetic telegraph (Ellison ca. 1950, 274).

41. See Wilcox 2007, 788; Yaszek 2005, 297–313.

42. Lippit's innovative reading of this scene as a reflection upon the use of atomic weapons in World War II (2005, 101–102) accords with this reading: it emphasizes how horrible violence can unintentionally provoke an experience of colorlessness or transcendent disembodiment.

43. Whitman's "Song of Myself" begins with *I* and ends with *you*, performing a distinctly American poetic act of expanding the narrative voice to speak for the many. By ending with this line, Ellison closes a similar circuit.

CONCLUSION

1. Such re-interpretations manifest in the use of terms such as *digital divide*. On social disparity and information, see Warschauer 2003, 11. On the discourse of this era as the "information age," see Kline 2015, 202–204. On the metaphor "genetic code," see Kay 2000. Donna Haraway (2004, 23–25) also illuminates this issue, arguing that "communications sciences and modern biologies are constructed by a common move—*the translation of the world into a problem of coding*."

2. In fact, Emily St. John Mandel's 2014 best-seller *Station Eleven* was a uniquely hopeful post-apocalyptic novel because it also imagined returning to electrical power a generation after a worldwide social and technological collapse.

3. As Edwards points out (2003, 193), the human casualties of so-called natural disasters are less likely to be caused by weather events themselves than by the effects of the weather on infrastructure.

4. The structuralist theorist Roman Jakobson (1971) argues that metonymy "underlies" literary realism—suggesting, perhaps, that this genre would lend itself to the study of metonymy. Jakobson also writes of metaphor and metonymy as "poles," constructing an electromagnetic analogy for understanding these two parts of speech.

5. See, for example, Oudshoorn and Pinch 2003; Kline and Pinch 1996.

6. Rosalind Williams discusses existential concerns at length in *The Triumph of Human Empire* (2013). See, for example, pp. 11–14.

REFERENCES

Adams, Henry. 1918. *The Education of Henry Adams* (reprint: Houghton Mifflin, 1973).

Adams, Henry. 1920. Rule of Phase Applied to History. In *The Degradation of the Democratic Dogma*, ed. Henry Adams and Brooks Adams. Macmillan.

Adams, Joseph H. 1907. *Harper's Electricity Book for Boys*. Harper.

Adas, Michael. 1990. *Machines as the Measure of Men: Science, Technology, and Ideologies of Western Dominance*. Cornell University Press.

Addams, Jane. 1912. *Twenty Years at Hull-House*. Macmillan.

Ahmad, Dohra. 2009. *Landscapes of Hope: Anti-Colonial Utopianism in America*. Oxford University Press.

Allen, Polly Wyn. 1988. *Building Domestic Liberty: Charlotte Perkins Gilman's Architectural Feminism*. University of Massachusetts Press.

Armstrong, Tim. 1998. *Modernism, Technology, and the Body: A Cultural Study*. Cambridge University Press.

Atherton, Gertrude. 1897. *Patience Sparhawk and Her Times* (reprint: Macmillan, 1908).

Atherton, Gertrude. 1932. *Adventures of a Novelist*. Liveright.

Auerbach, Jonathan. 1996. *Male Call: Becoming Jack London*. Duke University Press.

Baker, Houston A. 1983. To Move without Moving: An Analysis of Creativity and Commerce in Ralph Ellison's Trueblood Episode. *PMLA* 98 (5): 828–845.

Banner, Stuart. 2002. *The Death Penalty: An American History*. Harvard University Press.

Banta, Martha. 1991. The Boys and the Bosses: Twain's Double Take on Work, Play, and the Democratic Ideal. *American Literary History* 3 (fall): 487–520.

Barad, Karen. 2007. *Meeting the Universe Halfway: Quantum Physics and the Entanglement of Matter and Meaning.* Duke University Press.

Barrett, J. P. 1894. *Electricity at the Columbian Exposition.* Donnelley.

Barthes, Roland. 1986. The Reality Effect. In *The Rustle of Language.* Blackwell.

Baudrillard, Jean. 1994. *Simulacra and Simulation.* University of Michigan Press.

Bazerman, Charles. 1999. *The Languages of Edison's Light.* MIT Press.

Beard, George Miller. 1881. *American Nervousness: Its Causes and Consequences.* Putnam.

Beard, George Miller, and A. D. Rockwell. 1884. *A Practical Treatise on the Medical and Surgical Uses of Electricity.* William Wood.

Bellow, Saul. 1955. Dreiser and the Triumph of Art. In *The Stature of Theodore Dreiser*, ed. Alfred Kazin and Charles Shapiro. Indiana University Press.

Benjamin, Park. 1886. *The Age of Electricity: From Amber-Soul to Telephone.* Scribner.

Benjamin, Shanna Greene. 2014. There's Something About Mary: Female Wisdom and the Folk Presence in Ralph Ellison's *Invisible Man. Meridians* 12 (1): 121–148.

Bimber, Bruce. 1990. Karl Marx and the Three Faces of Technological Determinism. *Social Studies of Science* 20 (May): 333–351.

Bird, Arthur. 1899. *Looking Forward: A Dream of the United States of the Americas in 1999.* I. C. Childs.

Blair, Sara. 2005. Ralph Ellison, Photographer. *Raritan* 24 (spring): 21–44.

Boudreau, Kristin. 2006. *The Spectacle of Death: Populist Literary Responses To American Capital Cases.* Prometheus.

Brackett, Cyrus F., et al. 1891. *Electricity in Daily Life: A Popular Account of the Applications of Electricity to Every Day Uses.* Scribner.

Bradley, Adam. 2010. *Ralph Ellison in Progress: The Making and Unmaking of One Writer's Great American Novel.* Yale University Press.

Brandon, Craig. 1999. *The Electric Chair: An Unnatural American History.* McFarland.

Burnett, Frances Hodgson. 1895. *Two Little Pilgrims' Progress: A Story of the City Beautiful.* Scribner.

Cahan, David, and M. Eugene Rudd. 2001. *Science at the American Frontier: A Biography of DeWitt Bristol Brace.* University of Nebraska Press.

Callon, Michel. 2012. Society in the Making: The Study of Technology as a Tool for Sociological Analysis. In *The Social Construction of Technological Systems: New*

Directions in the Sociology and History of Technology, anniversary edition, ed. Wiebe E. Bijker, Thomas P. Hughes and Trevor Pinch. MIT Press.

Callon, Michel, Pierre Lascoumes, and Yannick Barthe. 2009. *Acting in an Uncertain World: An Essay on Technical Democracy*. MIT Press.

Camfield, Gregg. 1994. *Sentimental Twain: Samuel Clemens in the Maze of Moral Philosophy*. University of Pennsylvania Press.

Camus, Albert. 1946. *Neither Victims Nor Executioners* (reprint: Library Company of Pennsylvania, 1986).

Carey, James W. and John J. Quirk. 1970. The Mythos of the Electronic Revolution. *American Scholar* 39: 219–240 and 395–424.

Chase, Stuart. 1929. *Men and Machines*. Macmillan.

Clarke, Bruce. 2001. *Energy Forms: Allegory and Science in the Era of Classical Thermodynamics*. University of Michigan Press.

Conquergood, Dwight. 2002. Lethal Theatre: Performance, Punishment, and the Death Penalty. *Theatre Journal* 54 (October): 339–367.

Cowan, Ruth Schwartz. 1983. *More Work for Mother: The Ironies of Household Technology from the Open Hearth to the Microwave*. Basic Books.

Cowan, Ruth Schwartz. 2012. The Consumption Junction: A Proposal for Research Strategies in the Sociology of Technology. In *The Social Construction of Technological Systems: New Directions in the Sociology and History of Technology* anniversary edition, ed. Wiebe E. Bijker, Thomas P. Hughes, and Trevor Pinch. MIT Press.

Cox, James M. 1960. *A Connecticut Yankee in King Arthur's Court*: The Machinery of Self-Preservation. *Yale Review* 50 (autumn): 89–102.

Crane, Stephen. 1896. The Devil's Acre. *The World*. October 25:23.

Crowley, John W. 1999. *The Dean of American Letters: The Late Career of William Dean Howells*. University of Massachusetts Press.

Cummings, Sherwood. 1960. Mark Twain and the Sirens of Progress. *Journal of the Central Mississippi Valley American Studies Association* 1 (fall): 17–24.

Cunningham, Andrew, and Perry Williams, eds. 1992. *The Laboratory Revolution in Medicine*. Cambridge University Press.

Curtin, Maureen Frances. 2003. *Out of Touch: Skin Tropes and Identities in Woolf, Ellison, Pynchon, and Acker*. Routledge.

Daston, Lorraine, and Peter Galison. 2010. *Objectivity*. Zone Books.

Dauphin, Laurent. 1995. *The Valley of the Moon*: A Reassessment. In *The Critical Response to Jack London*, ed. Susan M. Nuernberg. Greenwood.

Delbourgo, James. 2006. *A Most Amazing Scene of Wonders: Electricity and Enlightenment in Early America*. Harvard University Press.

Dimock, Wai Chi. 1991. The Economy of Pain: Capitalism, Humanitarianism and the Realistic Novel. In *New Essays on the Rise of Silas Lapham*, ed. Donald Pease. Cambridge University Press.

Dimock, Wai Chi, and Ellen Spolsky. 1999. A Theory of Resonance. *PMLA* 114 (March): 221–223.

Dixon, Melvin. 1980. O, Mary Rambo, Don't You Weep. *Carleton Miscellany* 18 (fall): 98–104.

Dobski, Bernard J., Jr., and Benjamin Kleinerman. 2007. "We should see certain things yet, let us hope and believe": Technology, Sex, and Politics in Mark Twain's *Connecticut Yankee*. *Review of Politics* 69 (fall): 599–624.

Downey, Greg. 2001. Virtual Webs, Physical Technologies, Hidden Workers: The Spaces of Labor in Information Internetworks. *Technology and Culture* 42 (April): 209–235.

Dreiser, Theodore. 1900. *Sister Carrie* (reprint: Norton, 2006).

Dreiser, Theodore. 1925. *An American Tragedy* (reprint: Modern Library, 1953).

Driscoll, Kerry. 2004. "Man Factories" and the "White Indians" of Camelot: Re-reading the Native Subtext of *A Connecticut Yankee in King Arthur's Court*. *Mark Twain Annual* 2 (September): 7–23.

Eby, Clare Virginia. 1998. *Dreiser and Veblen: Saboteurs of the Status Quo*. University of Missouri Press.

Edgerton, David. 2006. *The Shock of the Old: Technological and Global History Since 1900*. Profile.

Edison, Thomas Alva. 1885. Electricity Man's Slave. *Scientific American* 52, March 21: 185.

Edwards, Paul N. 1997. *The Closed World: Computers and the Politics of Discourse in Cold War America*. MIT Press.

Edwards, Paul N. 2003. Infrastructure and Modernity: Force, Time, and Social Organization in the History of Sociotechnical Systems. In *Modernity and Technology*, ed. Thomas J. Misa, Philip Brey, and Andrew Feenberg. MIT Press.

Eglash, Ron. 2007. Broken Metaphor: The Master-Slave Analogy in Technical Literature. *Technology and Culture* 48 (April): 360–369.

Eglash, Ron, Jennifer Crossiant, Giovanna Di Chiro, and Rayvon Fouché, eds. 2004. *Appropriating Technology: Vernacular Science and Social Power*. University of Minnesota Press.

Electric Execution Act of 1888, Chapter 489 of the Laws of the State of New York.

References

Ellison, Ralph. 1944a. Flying Home. Reprinted in *Flying Home and Other Stories*, ed. John F. Callahan. Vintage International, 1998.

Ellison, Ralph. 1944b. King of the Bingo Game. Reprinted in *Flying Home and Other Stories*, ed. John F. Callahan. Vintage International, 1998.

Ellison, Ralph. circa 1950. Invisible Man manuscript variant. Ralph Ellison Papers, Library of Congress.

Ellison, Ralph. 1952. *Invisible Man* (reprint: Vintage International, 1980).

Ellison, Ralph. 1963. Out of the Hospital and Under the Bar. In *Soon, One Morning: New Writing by American Negroes, 1940–1962*, ed. Herbert Hill. Knopf.

Ellison, Ralph. 1964. *Shadow and Act* (reprint: Vintage International, 1995).

Ellison, Ralph. 2010. *Three Days Before the Shooting*. Random House.

Ellul, Jacques. 1964. *The Technological Society*. Vintage.

Essig, Mark. 2003. *Edison and the Electric Chair: A Story of Light and Death*. Walker.

Fleissner, Jennifer. 2004. *Women, Compulsion, Modernity: The Moment of American Naturalism*. University of Chicago Press.

Foley, Barbara. 2010. *Wrestling with the Left: The Making of Ralph Ellison's Invisible Man*. Duke University Press.

Folsom, James K. 1989. Imaginative Safety Valves: Frontier Themes in the Literature of the Gilded Age. In *The Frontier Experience and the American Dream: Essays on American Literature*, ed. David Mogen, Mark Bundy, and Paul Bryant. Texas A&M University Press.

Ford, Douglas. 1999. Crossroads and Cross-Currents in Invisible Man. *MFS: Modern Fiction Studies* 45 (winter): 887–904.

Fore, Devin. 2012. *Realism after Modernism: The Rehumanization of Art and Literature*. MIT Press.

Foucault, Michel. 1977. *Discipline and Punish: The Birth of the Prison*. Vintage.

Fouché, Rayvon. 2006. Say It Loud, I'm Black and I'm Proud: African Americans, African American Artifactual Culture, and Black Vernacular Technological Creativity. *American Quarterly* 58 (September): 639–661.

Franklin, Benjamin. 1751. *Experiments and Observations on Electricity* (reprint : E. Cave, 1761).

Fulton, Joe B. 1997. *Mark Twain's Ethical Realism: The Aesthetics of Race, Class, and Gender*. University of Missouri Press.

Fusco, Katherine. 2009. Systems, Not Men: Producing People in Charlotte Perkins Gilman's Herland. *Studies in the Novel* 41 (winter): 418–434.

Gair, Christopher Hugh. 1996. "The Way Our People Came": Citizenship, Capitalism, and Racial Difference in *The Valley of the Moon*. In *Rereading Jack*

London, ed. Leonard Cassuto and Jeanne Campbell Reesman. Stanford University Press.

Gamble, Vanessa Northington. 1997. Under the Shadow of Tuskegee: African Americans and Health Care. *American Journal of Public Health* 87 (November): 1773–1778.

Gannett, Henry. 1873. Electric Peak. *Sixth Annual Report of the United States Geological Survey of the Territories, Embracing Portions of Montana, Idaho, Wyoming, and Utah; Being a Report of Progress of the Explorations for the Year 1872*. Government Printing Office.

Gardiner, Jane. 1987. "A More Splendid Necromancy": Mark Twain's *Connecticut Yankee* and the Electrical Revolution. *Studies in the Novel* 19 (winter): 448–458.

Gates, Henry Louis, Jr. 1989. *The Signifying Monkey: A Theory of African-American Literary Criticism*. Oxford University Press.

Geddes, Patrick. 1915. *Cities in Evolution: An Introduction to the Town Planning Movement and to the Study of Civics*. Williams and Norgate.

Gerry, Elbridge T., Alfred P. Southwick, and Matthew Hale. 1888. *Report of the Commission to Investigate and Report the Most Humane and Practical Method of Carrying into Effect the Sentence of Death in Capital Cases*. Argus.

Gilman, Charlotte Perkins. 1901. *Concerning the Children*. Small, Maynard.

Gilman, Charlotte Perkins. 1903. *The Home: Its Work and Influence*. McClure, Phillips.

Gilman, Charlotte Perkins. 1904. *Human Work*. McClure, Phillips.

Gilman, Charlotte Perkins. 1908. Aunt Mary's Pie Plant. *Woman's Home Companion* 6 (June): 14, 48–49.

Gilman, Charlotte Perkins. 1911a. *The Man-Made World, or, Our Androcentric Culture*. Charlton.

Gilman, Charlotte Perkins. 1911b. *Moving the Mountain*. Charlton.

Gilman, Charlotte Perkins. 1991a. The Labor Movement. Reprinted in *Charlotte Perkins Gilman: A Nonfiction Reader*, ed. Larry Ceplair. Columbia University Press.

Gilman, Charlotte Perkins. 1991b. Socialist Psychology. Reprinted in *Charlotte Perkins Gilman: A Nonfiction Reader*, ed. Larry Ceplair. Columbia University Press.

Gilman, Charlotte Perkins. 1992. Dr. Clair's Place. Reprinted in *Herland and Selected Stories*, ed. Barbara H. Solomon. Signet Classics.

Gilman, Charlotte Perkins. 1998. *Women and Economics*. Dover.

Gilman, Charlotte Perkins. 1999. Bee Wise. Reprinted in *Herland, The Yellow Wall-Paper, and Selected Writings*, ed. Denise D. Knight. Penguin.

Gilman, Charlotte Perkins. 2012. *Charlotte Perkins Gilman's In This Our World and Uncollected Poems*, ed. Gary Scharnhorst and Denise D. Knight. Syracuse University Press.

Gilmore, Paul. 2008. *Aesthetic Materialism: Electricity and American Romanticism*. Stanford University Press.

Glazener, Nancy. 1997. *Reading For Realism: A History of a U.S. Literary Institution, 1850–1910*. Duke University Press.

Goble, Mark. 2010. *Beautiful Circuits: Modernism and the Mediated Life*. Columbia University Press.

Golden, Catherine J. 1999. "Written to Drive Nails With": Recalling the Early Poetry of Charlotte Perkins Gilman. In *Charlotte Perkins Gilman: Optimist Reformer*, ed. Jill Rudd and Val Gough. University of Iowa Press.

Goodling, Sara Britton. 2003. The Silent Partnership: Naturalism and Sentimentalism in the Novels of Rebecca Harding Davis and Elizabeth Stuart Phelps. In *Twisted from the Ordinary: Essays on American Literary Naturalism*, ed. Mary E. Papke. University of Tennessee Press.

Halliday, Sam. 2007. *Science and Technology in the Age of Hawthorne, Melville, Twain, and James: Thinking and Writing Electricity*. Palgrave Macmillan.

Haraway, Donna. 1990. *Simians, Cyborgs, and Women: The Reinvention of Nature*. Routledge.

Haraway, Donna. 2004. *The Haraway Reader*. Routledge.

Hård, Michael, and Andrew Jamison. 2005. *Hubris and Hybrids: A Cultural History of Technology and Science*. Routledge.

Harris, Neil. 1990. *Cultural Excursions: Marketing Appetites and Cultural Tastes in Modern America*. University of Chicago Press.

Haslam, Gerald. 1989. Golden State: A Demanding Paradise, an Enduring Frontier. In *The Frontier Experience and the American Dream: Essays on American Literature*, ed. David Mogen, Mark Busby and Paul Bryant. Texas A&M University Press.

Hayles, N. Katherine. 1999. *How We Became Posthuman: Virtual Bodies in Cybernetics, Literature, and Informatics*. University of Chicago Press.

Heidegger, Martin. 1977. *The Question Concerning Technology and Other Essays*. Harper & Row.

Hewett, John R. 1922. Is Democracy to Fail? *General Electric Review* 25 (June): 331–332.

Hilgartner, Stephen. 2015. Capturing the Imaginary: Vanguards, Visions and the Synthetic Biology Revolution. In *Science and Democracy: Making Knowledge and Making Power in the Biosciences and Beyond*, ed. Stephen Hilgartner, Clark A. Miller and Rob Hagendijk. Routledge.

Hill, Hamlin, ed. 1963. *A Connecticut Yankee in King Arthur's Court: A Facsimile of the First Edition.* Chandler.

Horkheimer, Max, and Theodor W. Adorno. 2002. *Dialectic of Enlightenment.* Stanford University Press.

Howells, William Dean. 1887. The Editor's Study. *Harpers New Monthly Magazine,* April. Reprinted in *Criticism and Fiction* (Harper, 1892).

Howells, William Dean. 1888. Execution by Electricity. *Harper's Weekly,* January 14, 23.

Howells, William Dean. 1907. *Through the Eye of a Needle: A Romance.* Harper.

Howells, William Dean. 1910. *My Mark Twain* (reprint: Louisiana State University Press, 1967).

Howells, William Dean, and Mark Twain. 1960. Colonel Sellers as a Scientist. In *The Complete Plays of W. D. Howells,* ed. Walter J. Meserve, William M. Gibson, and George Arms. New York University Press.

Hughes, Thomas P. 1979. The Electrification of America: The System Builders. *Technology and Culture* 20 (January): 124–161.

Hughes, Thomas P. 1983. *Networks of Power: Electrification in Western Society, 1880–1930.* Johns Hopkins University Press.

Hughes, Thomas P. 1988. The Industrial Revolution That Never Came. *American Heritage of Invention & Technology* 3 (winter): 58–64.

Hughes, Thomas P. 2004. *American Genesis: A Century of Invention and Technological Enthusiasm, 1870–1970.* University of Chicago Press.

Hughes, Thomas P., and Agatha C. Hughes. 1990. General Introduction: Mumford's Modern World. In *Lewis Mumford: Public Intellectual,* ed. Thomas P. Hughes and Agatha C. Hughes. Oxford University Press.

Inge, Thomas, ed. 2008. *A Connecticut Yankee in King Arthur's Court.* Oxford University Press.

Jakle, John A. 2001. *City Lights: Illuminating the American Night.* Johns Hopkins University Press.

Jakobson, Roman. 1971. *Selected Writings.* vol. 2. Mouton.

James, George Wharton. 1906. *The Wonders of the Colorado Desert.* Little, Brown.

Jameson, Frederic. 2005. *Archaeologies of the Future: The Desire Called Utopia and Other Science Fictions.* Verso.

Jasanoff, Sheila. 2015. Future Imperfect: Science, Technology, and the Imaginations of Modernity. In *Dreamscapes of Modernity: Sociotechnical Imaginaries and the Fabrication of Power,* ed. Sheila Jasanoff and Sang-Hyun Kim. University of Chicago Press.

Jasanoff, Sheila, and Sang-Hyun Kim. 2009. Containing the Atom: Sociotechnical Imaginaries and Nuclear Power in the United States and South Korea. *Minerva* 47 (June): 119–146.

Johns, Jillian. 2007. Jim Trueblood and His Critic-Readers: Ralph Ellison's Rhetoric of Dramatic Irony and Tall Humor in the Mid-Century American Literary Public Sphere. *Texas Studies in Literature and Language* 49 (September): 230–264.

Jordan, John M. 1989. "Society Improved the Way You Can Improve a Dynamo": Charles P. Steinmetz and the Politics of Efficiency. *Technology and Culture* 30 (January): 57–82.

Juma, Calestous. 2016. *Innovation and Its Enemies: Why People Resist New Technologies*. Oxford University Press.

Kasson, John F. 1999. *Civilizing the Machine: Technology and Republican Values in America, 1776–1900*. Hill and Wang.

Kay, Lily. 2000. *Who Wrote the Book of Life?: A History of the Genetic Code*. Stanford University Press.

Kline, Ronald R. 1992. *Steinmetz: Engineer and Socialist*. Johns Hopkins University Press.

Kline, Ronald R. 1995. Construing "Technology" as "Applied Science": Public Rhetoric of Scientists and Engineers in the United States, 1880–1945. *Isis* 86 (June): 194–221.

Kline, Ronald R. 2015. *The Cybernetics Moment, or, Why We Call Our Age the Information Age*. Johns Hopkins University Press.

Kline, Ronald R., and Trevor Pinch. 1996. Users as Agents of Technological Change: The Social Construction of the Automobile in the Rural United States. *Technology and Culture* 37 (October): 763–795.

Knoper, Randall. 1995. *Acting Naturally: Mark Twain in the Culture of Performance*. University of California Press.

Knoper, Randall. 2002. American Literary Realism and Nervous "Reflexion." *American Literature* 74 (December): 715–745.

Lakoff, George, and Mark Johnson. 1980. *Metaphors We Live By*. University of Chicago Press.

Latour, Bruno. 1996. Do Scientific Objects Have a History?: Pasteur and Whitehead in a Bath of Lactic Acid. *Common Knowledge* 5 (spring): 76–91.

Latour, Bruno, and Steve Woolgar. 1986. *Laboratory Life: The Construction of Scientific Facts*. Princeton University Press.

La Vergata, Antonello. 1994. Herbert Spencer: Biology, Sociology, and Cosmic Evolution. In *Biology as Society, Society as Biology: Metaphors*, ed. Peter Weingart, Sabine Maasen and Everett Mendelsohn. Kluwer.

Lawlor, Mary. 2000. *Recalling the Wild: Naturalism and the Closing of the American West*. Rutgers University Press.

Layton, Edwin T., Jr. 1971. *The Revolt of the Engineers: Social Responsibility and the American Engineering Profession*. Press of Case Western Reserve University.

Lears, T. J. Jackson. 1983. *No Place of Grace: Antimodernism and the Transformation of American Culture*. University of Chicago Press.

Lee, Carol A. 1989. Wired Help for the Farm Individual Electric Generating Sets for Farms, 1880–1930. PhD dissertation, Pennsylvania State University.

Leider, Emily Wortis. 1993. *California's Daughter: Gertrude Atherton and Her Times*. Stanford University Press.

Lenoir, Timothy. 1994. Helmholtz and the Materialities of Communication. *Osiris* 9:185–207.

Lenoir, Timothy, ed. 1998. *Inscribing Science: Scientific Texts and the Materiality of Communication*. Stanford University Press.

Lerer, Seth. 2003. Hello, Dude: Philology, Performance, and Technology in Mark Twain's *Connecticut Yankee*. *American Literary History* 15 (fall): 479–481.

Lieberman, Jennifer. 2010. Hank Morgan's Power Play: Electrical Networks in King Arthur's Court. *Mark Twain Annual* 8: 61–75.

Lieberman, Jennifer. 2015. Ralph Ellison's Technological Humanism. *MELUS* 40 (4): 8–27.

Lieberman, Jennifer. 2016. The Myth of the First African-American Electrical Engineer: Arthur U. Craig and the Importance of Teaching in Technological History. *History and Technology* 32 (May): 70–90.

Lieberman, Jennifer, and Ronald Kline. 2017. Dream of an Unfettered Electrical Future: Nikola Tesla, the Electrical Utopian Novel, and an Alternative American Sociotechnical Imaginary. *Configurations* 25 (1): 1–27.

Lippit, Akira Mizuta. 2005. *Atomic Light (Shadow Optics)*. University of Minnesota Press.

London, Jack. 1908. *Martin Eden*. Review of Reviews Company.

London, Jack. 1910. *Burning Daylight*. Macmillan.

London, Jack. 1913a. *John Barleycorn*. Century.

London, Jack. 1913b. *The Valley of the Moon*. Macmillan.

London, Jack. 1914. The Dream of Debs. In *The Strength of the Strong*. Macmillan.

London, Jack. 1988. *The Letters of Jack London*. Stanford University Press.

Lubar, Steven. 1998. Men/Women/Production/Consumption. In *His and Hers: Gender, Consumption, and Technology*, ed. Roger Horowitz and Arwen Mohun. University Press of Virginia.

Luhmann, Niklas. 1990. *Essays on Self-Reference*. Columbia University Press.

Lukács, Georg. 1963. *The Meaning of Contemporary Realism*. Merlin.

Lutes, Jean Marie. 2007. *Front Page Girls: Women Journalists in American Culture and Fiction, 1880–1930*. Cornell University Press.

MacDonald, Carlos D. 1890. Report of Carlos F. MacDonald on the Execution by Electricity of William Kemmler, Alias John Hart. Presented to the Governor September 20, 1890.

MacDougall, Robert. 2013. *The People's Network: The Political Economy of the Telephone in the Gilded Age*. University of Pennsylvania Press.

MacKenzie, Donald. 1996. *Knowing Machines: Essays on Technical Change*. MIT Press.

Malina, Roger F. 2009. Intimate Science and Hard Humanities. *Leonardo* 42 (May): 184.

Marchand, Roland, and Michael L. Smith. 1997. Corporate Science on Display. In *Scientific Authority and Twentieth-Century America*, ed. Ronald G. Walters. Johns Hopkins University Press.

Markus, Thomas A. 2002. *Embodied Utopias: Gender, Social Change, and the Modern Metropolis*. Routledge.

Marshall, Edward. 1912. The Woman of the Future: An Interview with Thomas Edison. Reprinted in *The Good Housekeeping Treasury: Selected from the Complete File* (Simon and Schuster, 1960).

Marvin, Carolyn. 1988. *When Old Technologies Were New: Thinking About Electric Communication in the Late Nineteenth Century*. Oxford University Press.

Marvin, Thomas F. 1996. Children of Legba: Musicians at the Crossroads in Ralph Ellison's *Invisible Man*. *American Literature* 68 (September): 587–608.

Marx, Leo. 1964. *The Machine in the Garden: Technology and the Pastoral Ideal in America*. Oxford University Press.

Marx, Leo. 1988. *The Pilot and the Passenger: Essays on Literature, Technology, and Culture in the United States*. Oxford University Press.

Marx, Leo. 1990. Lewis Mumford: Prophet of Organicism. In *Lewis Mumford: Public Intellectual*, ed. Thomas P. Hughes and Agatha C. Hughes. Oxford University Press.

Marx, Leo. 1994. The Idea of "Technology" and Postmodern Pessimism. In *Does Technology Drive History? The Dilemma of Technological Determinism*, ed. Merritt Roe Smith and Leo Marx. MIT Press.

Marx, Leo. 1997. *Technology*: The Emergence of a Hazardous Concept. *Social Research* 64 (fall): 965–988.

Marx, Leo, and Merritt Roe Smith, eds. 1994. *Does Technology Drive History? The Dilemma of Technological Determinism.* MIT Press.

McGhie, John. 1896. Long-Distance Transmission of Power by Electricity in the United States. *Cassier's Magazine* 9 (February): 359–374.

Meister, Morris. 1930. *Magnetism and Electricity.* Scribner.

Merton, Robert K. 1968. *Social Theory and Social Structure.* Free Press.

Michaels, Walter Benn. 1987. *The Gold Standard and the Logic of Naturalism.* University of California Press.

Michelson, Bruce. 1991. Realism, Romance, and Dynamite: The Quarrel of *A Connecticut Yankee in King Arthur's Court. New England Quarterly* 64 (December): 609–632.

Michelson, Bruce. 2006. *Printer's Devil: Mark Twain and the American Publishing Revolution.* University of California Press.

Mirowski, Philip. 1991. *More Heat Than Light: Economics as Social Physics, Physics as Nature's Economics.* Cambridge University Press.

Mirowski, Philip. 2004. *Effortless Economy of Science.* Duke University Press.

Misa, Thomas J. 2003. The Compelling Tangle of Modernity and Technology. In *Modernity and Technology,* ed. Thomas J. Misa, Philip Brey, and Andrew Feenberg. MIT Press.

Misa, Thomas J. 2004. *Leonardo to the Internet: Technology and Culture from the Renaissance to the Present.* Johns Hopkins University Press.

Molella, Arthur P. 1990. Mumford in Historiographical Context. In *Lewis Mumford: Public Intellectual,* ed. Thomas P. Hughes and Agatha C. Hughes. Oxford University Press.

Moran, Richard. 2003. *Executioner's Current: Thomas Edison, George Westinghouse, and the Invention of the Electric Chair.* Vintage.

Mumford, Lewis. 1926. *The Golden Day: A Study in American Literature and Culture* (reprint: Dover, 1968).

Mumford, Lewis. 1930. The Drama of Machines. *Scribner's Magazine,* August: 150–160.

Mumford, Lewis. 1934. *Technics and Civilization* (reprint: University of Chicago Press, 2010).

Nadel, Alan. 1991. *Invisible Criticism: Ralph Ellison and the American Canon.* University of Iowa Press.

Naito, Jonathan Tadashi. 2006. Cruel and Unusual Light: Electricity and Effacement in Stephen Crane's *The Monster. Arizona Quarterly* 62 (spring): 35–63.

National Police Gazette. 1890. Electrocuted. Volume 6, August 23: 23.

Nye, David E. 1985. *Image Worlds: Corporate Identities at General Electric, 1890–1930*. MIT Press.

Nye, David E. 1992. *Electrifying America: Social Meanings of a New Technology, 1880–1940*. MIT Press.

Nye, David E. 1994. *American Technological Sublime*. MIT Press.

Nye, David E. 2003. *America as Second Creation: Technology and Narratives of New Beginnings*. MIT Press.

Nye, David E. 2007. *Technology Matters: Questions to Live With*. MIT Press.

Nye, David E. 2010. *When the Lights Went Out: A History of Blackouts in America*. MIT Press.

Oldenziel, Ruth. 1999. *Making Technology Masculine: Men, Women, and Modern Machines in America, 1870–1945*. Amsterdam University Press.

Otis, Laura. 2001. *Networking: Communicating with Bodies and Machines in the Nineteenth Century*. University of Michigan Press.

Otis, Laura. 2002. The Metaphoric Circuit: Organic and Technological Communication in the Nineteenth Century. *Journal of the History of Ideas* 63 (January): 105–128.

Oudshoorn, Nelly, and Trevor Pinch, eds. 2003. *How Users Matter: The Co-Construction of Users and Technology*. MIT Press.

Pfaelzer, Jean. 1984. *The Utopian Novel in America, 1886–1896: The Politics of Form*. University of Pittsburgh Press.

Pfitzer, Gregory M. 1994. "Iron Dudes and White Savages in Camelot": The Influence of Dime-Novel Sensationalism on Twain's *A Connecticut Yankee in King Arthur's Court*. *American Literary Realism* 27 (fall): 42–58.

Pinch, Trevor J., and Wiebe E. Bijker. 2012. The Social Construction of Facts and Artifacts: Or How the Sociology of Science and the Sociology of Technology Might Benefit Each Other. In *The Social Construction of Technological Systems: New Directions in the Sociology and History of Technology*, anniversary edition, ed. Wiebe E. Bijker, Thomas P. Hughes and Trevor Pinch. MIT Press.

Pizer, Donald, ed. 2011. *Theodore Dreiser's An American Tragedy: A Documentary Volume*. Gale Cengage.

Plotnick, Rachel. 2012. At the Interface: The Case of the Electric Push Button, 1880–1923. *Technology and Culture* 53 (October): 815–845.

Poovey, Mary. 2001. The Model System of Contemporary Literary Criticism. *Critical Inquiry* 27 (spring): 408–438.

Pope, Franklin Leonard. 1889. *Evolution of the Electric Incandescent Lamp*. Cook and Hall.

Porter, Theodore M. 1995. *Trust in Numbers: The Pursuit of Objectivity in Science and Public Life*. Princeton University Press.

Pursell, Carol. 1995. *The Machine in America: A Social History of Technology*. Johns Hopkins University Press.

Rabinovitz, Lauren. 2012. *Electric Dreamland: Amusement Parks, Movies, and American Modernity*. Columbia University Press.

Ragin, Charles C., and Howard S. Becker. 1992. *What Is A Case? Exploring the Foundations of Social Inquiry*. Cambridge University Press.

Reading List for Democracy. 1942. *Journal of Educational Sociology* 15 (March): 430–441.

Rescher, Nicholas. 1996. *Priceless Knowledge? Natural Science in Economic Perspective*. Rowman & Littlefield.

Rigal, Laura. 2003. Imperial Attractions: Franklin's New Experiments of 1751. In *Memory Bytes: Historical Perspectives on American Digital Culture*, ed. Lauren Rabinovitz and Abraham Geil. Duke University Press.

Riskin, Jessica. 1999. The Lawyer and the Lightning Rod. *Science in Context* 12: 61–99.

Rockwell, Alphonso David. 1920. *Rambling Recollections: An Autobiography*. Paul B. Hoeber.

Roemer, Kenneth. 1976. *The Obsolete Necessity: America in Utopian Writings, 1888–1900*. Kent State University Press.

Roggenkamp, Karen. 2005. *Narrating the News: New Journalism And Literary Genre in Late Nineteenth-Century American Newspapers and Fiction*. Kent State University Press.

Rowe, John Carlos. 1995. How the Boss Played the Game: Twain's Critique of Imperialism in *A Connecticut Yankee in King Arthur's Court*. In *The Cambridge Companion to Mark Twain*, ed. Forrest G. Robinson. Cambridge University Press.

Ryan, Ann M. 2009. The Voice of Her Laughter: Mark Twain's Tragic Feminism. *American Literary Realism* 41 (spring): 192–213.

Sachs, Aaron. 2007. *The Humboldt Current: A European Explorer and His American Disciples*. Oxford University Press.

Sarat, Austin. 2001. *When the State Kills: Capital Punishment and the American Condition*. Princeton University Press.

Schatzberg, Eric. 2001. Culture and Technology in the City: Opposition to Mechanized Street Transportation in Late-Nineteenth-Century America. In *Technologies of Power: Essays in Honor of Thomas Parke Hughes and Agatha Chipley Hughes*, ed. Michael Thad Allen and Gabrielle Hecht. MIT Press.

Schatzberg, Eric. 2006. *Technik* Comes to America: Changing Meanings of *Technology* before 1930. *Technology and Culture* 47 (July): 486–512.

Schivelbusch, Wolfgang. 1988. *Disenchanted Night: The Industrialization of Light in the Nineteenth Century*. University of California Press.

Scientific American. 1876. Electricity as an Executioner. Volume 34, no. 2: 15.

Scientific American. 1885. Electricity for Executing Criminals. Volume 52, no. 7: 101.

Scientific American. 1889. Death by Electricity—the New Law of New York. Volume 60, no. 1: 2.

Schwed, Roger. 1983. *Abolition and Capital Punishment*. AMS Press.

Segal, Howard P. 2005. *Technological Utopianism in American Culture*, twentieth anniversary edition. Syracuse University Press.

Seltzer, Mark. 1992. *Bodies and Machines*. Routledge.

Sewell, David R. 1994. Hank Morgan and the Colonization of Utopia. In *Mark Twain: A Collection of Critical Essays*, ed. Eric Sundquist. Prentice-Hall.

Shanley, Mary Lyndon, and Peter G. Stillman. 1982. Mark Twain: Technology, Social Change, and Political Power. In *The Artist and Political Vision*, ed. Benjamin R. Barber and Michael J. Gargas McGrath. Transaction.

Shaw, George M. 1878. "Sketch of Thomas Alva Edison." *Popular Science Monthly* (August): 487–491.

Shulman, Robert. 1987. The War Machine in the Garden: Capitalism, Republicanism, and Protestant Character Structure in *A Connecticut Yankee*. In *Social Criticism and Nineteenth-Century American Fictions*. University of Missouri Press.

Siskin, Clifford. 2016. *System: The Shaping of Modern Knowledge*. MIT Press.

Slaton, Amy E. 2010. *Race, Rigor, and Selectivity in U.S. Engineering: The History of an Occupational Color Line*. Harvard University Press.

Slotkin, Richard. 1994. Mark Twain's Frontier, Hank Morgan's Last Stand. In *Mark Twain: A Collection of Critical Essays*, ed. Eric Sundquist. Prentice-Hall.

Smith, Henry Nash. 1964a. *Mark Twain's Fable of Progress: Political and Economic Ideas in A Connecticut Yankee*. Rutgers University Press.

Smith, Henry Nash. 1964b The Search for a Capitalist Hero: Businessmen in American Fiction. In *The Business Establishment*, ed. Earl F. Cheit. Wiley.

Smith, Michael L. 1994. Recourse of Empire: Landscapes of Progress in Technological America. In *Does Technology Drive History? The Dilemma of Technological Determinism*, ed. Merritt Roe Smith and Leo Marx. MIT Press.

Solnit, Rebecca. 2013. *The Faraway Nearby*. Penguin.

Stallman, Robert Wooster. 1952. Introduction. In *Stephen Crane: Stories and Tales*. Vintage.

Steele, James W. 1895. *Steam, Steel and Electricity*. Werner.

Steinmetz, Charles P. 1913. The Future Development of the Electrical Business. In *Camp Co-operation: Book of Proceedings*. Association Island.

Steinmetz, Charles P. 1916. *America and the New Epoch*. Harper.

Stepan, Nancy Leys. 1986. Race and Gender: The Role of Analogy in Science. *Isis* 77 (June): 261–277.

Suchman, Lucy. 2007. *Human-Machine Reconfigurations: Plans and Situated Actions*, second edition. Cambridge University Press.

Tate, Claudia. 1987. Notes on the Invisible Women in Ralph Ellison's *Invisible Man*. In *Speaking for You: The Vision of Ralph Ellison*, ed. Kimberly W. Benston. Howard University Press.

Terman, Frederick E. 2002. A Brief History of Electrical Engineering Education. *Proceedings of the IEEE* 86 (August): 1792–1800.

Tesla, Nikola. 1900. The Problem of Increasing Human Energy. *Century Magazine* (June): 175–210.

The Electrical World. 1890. The Execution of Kemmler by Electricity. Volume 16, no. 7: 99–100.

Thomas de la Peña, Carolyn 2003. *The Body Electric: How Strange Machines Built the Modern American*. NYU Press.

Thompson, Carl D. 1932. *Confessions of the Power Trust*. Dutton.

Thrailkill, Jane. 2007. *Affecting Fictions: Mind, Body, and Emotion in American Literary Realism*. Harvard University Press.

Tichi, Cecilia. 1987. *Shifting Gears: Technology, Literature, and Culture in Modernist America*. University of North Carolina Press.

Toomer, Jean. 1923. *Cane*. Boni & Liveright.

Trachtenberg, Alan. 1982. *The Incorporation of America: Culture and Society in the Gilded Age*. Hill and Wang.

Tresch, John. 2012. *The Romantic Machine: Utopian Science and Technology after Napoleon*. University of Chicago Press.

Trowbridge, John. 1891. *The Electrical Boy, or the Career of Greatman and Greatthings*. Little, Brown.

Turkle, Sherry. 2011. *Alone Together: Why We Expect More from Technology and Less from Each Other*. Basic Books.

Turner, Frederick Jackson. 1921. *The Frontier in American History*. Holt.

Tuttle, Jennifer S. 2000. Rewriting the West Cure: Charlotte Perkins Gilman, Owen Wister, and the Sexual Politics of Neurasthenia. In *The Mixed Legacy of Charlotte Perkins Gilman*, ed. Catherin J. Golden and Joanna Schneider Zangrando. University of Delaware Press.

Twain, Mark. 1889. *A Connecticut Yankee in King Arthur's Court* (reprint: Oxford University Press, 1996).

Twain, Mark. 1899. Mrs. McWilliams and the Lightning. In *The American Claimant and Other Stories and Sketches*. Harper.

Twain, Mark. 1940. *Mark Twain's Travels with Mr. Brown: being heretofore uncollected sketches written by Mark Twain for the San Francisco Alta California in 1866 & 1867, describing the adventures of the author and his irrepressible companion in Nicaragua, Hannibal, New York, and other spots on their way to Europe*. Knopf.

Twain, Mark. 1979. *Mark Twain's Notebooks and Journals*, volume III, ed. Robert Pack Browning, Michael B. Frank, and Lin Salama. University of California Press.

Underwood, Doug. 2013. *The Undeclared War Between Journalism and Fiction: Journalists as Genre Benders in Literary History*. Palgrave Macmillan.

Veblen, Thorstein B. 1898. Why Is Economics Not an Evolutionary Science? *Quarterly Journal of Economics* 12 (3): 373–397.

Veblen, Thorstein B. 1912. *Theory of the Leisure Class: An Economic Study of Institutions*. Macmillan.

Veblen, Thorstein B. 1921. *The Engineers and the Price System*. B. W. Huebsch.

Vila, Bryan, and Cynthia Morris, eds. 1997. The Abolition Movement. In *Capital Punishment in the United States: A Documentary History*. Greenwood.

Wajcman, Judy. 1991. *Feminism Confronts Technology*. Pennsylvania State University Press.

Wandler, Steven. 2010. Hogs, Not Maidens: The Ambivalent Imperialism of *A Connecticut Yankee in King Arthur's Court*. *Arizona Quarterly* 66 (winter): 33–52.

Ward, Lester Frank. 1893. *The Psychic Factors of Civilization*. Ginn.

Warschauer, Mark. 2003. *Technology and Social Inclusion: Rethinking the Digital Divide*. MIT Press.

Wellman, Barry. 2002. Little Boxes, Glocalization, and Networked Individualism. In *Digital Cities II: Computational and Sociological Approaches*, ed. Makoto Tanabe, Peter van den Besselaar, and Toru Ishida. Springer.

Wells-Barnett, Ida, ed. 1893. *The Reason Why the Colored American Is Not in the World's Columbian Exposition* (reprint: University of Illinois Press, 1999).

White, Hayden. 1973. *Metahistory: The Historical Imagination in Nineteenth-Century Europe*. Johns Hopkins University Press.

Whitman, Walt. 1855. *Leaves of Grass* (reprint, ed. Carl Sandburg: Modern Library, 1921).

Wilcox, Johnnie. 2007. Black Power: Minstrelsy and Electricity in Ralph Ellison's *Invisible Man*. *Callaloo* 30 (fall): 987–1009.

Williams, Raymond. 1977. *Marxism and Literature*. Oxford University Press.

Williams, Rosalind. 1990a. *Notes on the Underground: An Essay on Technology, Society, and the Imagination*. MIT Press.

Williams, Rosalind. 1990b. Lewis Mumford as Historian of Technology in *Technics and Civilization*. In *Lewis Mumford: Public Intellectual*, ed. Thomas P. Hughes and Agatha C. Hughes. Oxford University Press.

Williams, Rosalind. 2013. *The Triumph of Human Empire*. University of Chicago Press.

Wimsatt, William K., and Monroe C. Beardsley. 1946. The Intentional Fallacy. *Sewanee Review* 54: 468–488.

Winner, Langdon. 1986. Do Artifacts Have Politics? In *The Whale and the Reactor: a Search for Limits in an Age of High Technology*. University of Chicago Press.

Woolgar, Steve. 2012. Reconstructing Man and Machine: A Note on Sociological Critiques of Cognitivism. In *The Social Construction of Technological Systems: New Directions in the Sociology and History of Technology*, anniversary edition, ed. Wiebe E. Bijker, Thomas P. Hughes and Trevor Pinch. MIT Press.

Wright, John S. 2003. "Jack-the-Bear" Dreaming: Ellison's Spiritual Technologies. *boundary 2* 30 (summer): 175–194.

Wright, John S. 2005. Ellison's Experimental Attitude and the Technologies of Illumination. In *The Cambridge Companion to Ralph Ellison*, ed. Ross Posnock. Cambridge University Press.

Wright, John S. 2006. *Shadowing Ralph Ellison*. University Press of Mississippi.

Yaszek, Lisa. 2005. An Afrofuturist Reading of Ralph Ellison's *Invisible Man*. *Rethinking History* 9 (2/3): 297–313.

Youngberg, Quentin. 2005. Morphology of Manifest Destiny: The Justified Violence of John O'Sullivan, Hank Morgan, and George W. Bush. *Canadian Review of American Studies* 35 (fall): 315–333.

INDEX

Adams, Henry, 10–12, 191, 212
Addams, Jane, 140
Adorno, Theodor, 28, 235n14
Advertising, 38, 111, 125, 136, 151, 154, 176–183
Alternating current, 55, 69, 151–152
Anti-modernism, 11, 87, 164, 172, 184, 216
Armstrong, Tim, 93
Atherton, Gertrude, 14, 57–58, 66, 74–83, 86–90, 217
Ayrton, Hertha, 103, 236n16

Barad, Karen, 235n14, 238n28
Battery, electric
  as artifact, 14, 31–32, 53, 105–107
  as metaphor or analogy, 14, 97–109, 111, 113, 117, 122
Battle of the Currents, 54–56
Bazerman, Charles, 136
Beard, George M., 14, 110–112, 117, 122, 124, 142
Bellamy, Edward, 93, 95–96, 114–115
"Body electric," idea of, 5, 93, 98–99, 105, 110–112, 124, 183. *See also* Electricity, and the human body; Nerves
Button, electric
  as an artifact, 2, 24, 29, 32
  and electric execution, 61–63
  as a symbol, 24, 31–33, 41, 43, 90, 198, 214

Capitalism, 38, 60, 132–133, 143–163
Capital punishment. *See* Death penalty; Electric execution; Electric Execution Act
Carey, James W., 125
Chicago Columbian Exposition, 90, 121
Class, 31, 38, 60, 77, 103–104, 114, 121, 132–139, 155, 181
  tensions exacerbated by electricity, 31, 121, 132, 155
  tensions mitigated by electricity, 38, 114, 132–136
Clemens, Samuel. *See* Twain, Mark
Colonization, 45, 156, 158, 160
Communication systems, 3–4, 17, 24, 31, 40, 47, 104, 111, 161–162
Consumers, 7, 15, 102–103, 120, 148–150, 161–162, 213
  and choice, 161–162
  the erasure of, 148–149
  as feminine, 102–103, 120, 150, 236n22
Cowan, Ruth Schwartz, 103, 121, 148, 176, 236n18
Crane, Stephen, 14, 63–66, 71, 79, 84, 169

Davis, Edwin F., 68, 73, 233n26
Death penalty, 51–54, 57–62, 67, 75, 88. *See also* Electric execution; Electric Execution Act
  and the abolition movement, 52, 55, 68, 88

Delbourgo, James, 28–29
Democratization. *See* Electricity, as democratizing
Determinism, 38–39, 44, 145, 147, 169–172, 181, 201, 206
Dime novels, 69–70, 96, 131, 135–136, 139, 146
Dimock, Wai Chi, 45
Direct current, 55, 112, 151
Dreiser, Theodore, 14, 57–58, 60, 75–91, 108, 125, 144–145, 193, 217
 An American Tragedy, 57–58, 75–90
 Sister Carrie, 1, 45, 233n34

Edison, Thomas, 6, 54–55, 102–104, 112, 119–121, 136, 144
 biases about women and minorities, 102–104, 119–121
 and electric execution, 54–55
 as a symbol, 104, 136, 151, 186, 191
Edisonade. *See* Dime novels
Edwards, Paul, 4, 9, 11, 215
Electrical industry, 15, 38, 95, 102–105, 127, 136–140, 150–155, 163, 176, 181, 215. *See also* Advertising; General Electric
 promotional materials for, 136–140, 150–155, 176, 181
Electric execution. *See also* Death penalty
 and the "Battle of the Currents," 54–55
 as compared to hanging, 51–56, 59–60, 66, 68, 71–72
 and the electric chair, 5, 14, 52, 54–59, 63–90, 106, 199–200, 217
 and the fantasy of instantaneousness, 14, 53–54, 56–70, 83, 87, 106
 legality of, 55–56, 67, 69
 as scientific and perfectible, 54, 59–62, 68–73
 as torturous, 63–66, 85
 as unreal, 71–72, 79, 84, 86–87
Electric Execution Act, 55–56, 63, 66

Electricity. *See also* Battery; Button; Electrical industry; Water power
 comparisons to (analogies, metaphors, and metonymies), 7–10, 12, 20–21, 25–26, 36, 48, 66, 107, 110–111, 156, 189–191, 200, 205, 211 (*see also* Genre; Metaphors; Metonymy)
 as a concept, 4–5, 7, 24, 35, 42, 60, 93, 164–165, 175–184, 211–212
 as dangerous/fatal, 26, 29, 42–43, 52–53, 56, 61, 66, 147, 183–184, 199–203
 as democratizing, 31, 45, 48, 125, 170, 185, 211
 and the environment, 116–118, 125, 141–142
 and fantasies of control, 4, 21, 25–39, 42, 62, 89–90, 91, 96, 124–126, 146, 150, 159, 184, 191, 205, 213–215, 218
 generation of, 53, 55, 66–67, 99, 104, 117–119, 132–133, 150, 153–154, 162, 164, 180, 188, 215–216
 and the human body, 25, 43, 55, 68, 93–94, 98–99, 101, 105, 110–112, 117, 122, 124, 142, 183, 184, 199–201, 213, 219
 as interconnected, 5, 7, 12, 41, 43, 47, 95–96, 98–99, 104, 112–113, 124, 127
 and irrigation, 119, 132, 156, 158–159
 as life force, 1, 4, 7, 24, 43, 56, 89, 92, 111, 113, 183, 200, 219
 and medicine, 5, 49, 54, 93, 109–112, 117, 123, 200–201, 219
 as modern/progressive, 3, 5, 10, 46, 92, 101, 107, 158, 164–165, 175, 181, 188–191, 201, 204–205
 as natural, 4, 26–27, 29, 34, 66, 93, 119, 141–142, 146, 183, 219
 as normalized or naturalized, 22, 139–140, 149, 150–151, 153, 175
 as plural and ambiguous, 4–5, 13, 19, 24, 27, 29, 34–35, 41, 43, 74, 90, 164–165, 183, 199, 200, 211

Index

as safe and useful, 27, 29, 37, 61, 126–127, 132, 150, 165
as a slave or servant, 37–40, 174–183 (*see also* Slavery)
and social power, 1, 46–48, 93, 102–103, 119, 130, 148, 155, 158–160, 183–188, 191, 202–206
transmission of, 15, 104–107, 112, 119, 133–134, 137–140, 151–155, 159, 162, 180, 215, 217
and the "unplug" metaphor, 204–205, 213
and utopianism, 91, 95–96, 113–128, 132, 152, 175–183, 186, 196
Electrification
and the debate about private ownership, 126–127, 137, 153, 161
and rural America, 98, 154, 162, 178, 180, 186, 240n14
of the West, 138–139, 146–155
Electrocution, accidental, 52–54. *See also* Electric execution
Ellison, Ralph, 1, 15, 167–175, 183–209, 215, 217–218
Invisible Man, 1, 168–175, 183–209
"Battle Royal" episode, 183–187
and "The Golden Day," 192–197, 207
manuscript variant of the "Out of the Hospital" episode, 198–206
"Out of the Hospital" episode, 197–206
"Trueblood" episode, 187–192
Ellul, Jacques, 225n6
Emerson, Ralph Waldo, 169, 193, 207
Engineering, as a profession, 84, 120, 131, 139, 144–145, 151–152, 158, 181. *See also* Gender, and the presumed masculinity of expertise
as represented by the electrical industry, 137–139, 151–155

Fleissner, Jennifer, 10, 93, 149
Fouché, Rayvon, 187, 204, 244n25

Franklin, Benjamin, 11, 28–30, 37, 58–59, 104, 186
Frontier, idea of, 118, 133, 138–143, 146, 160

Galvani, Luigi, 26
Gender, 91–92, 101–103, 120–121, 136, 143, 150, 172, 191
and the presumed masculinity of expertise, 103, 120, 136, 172, 191
General Electric, 95–96, 103, 137, 151, 176–178, 181–183. *See also* Advertising; Electrical industry
Genre, hybridizations of, 16, 31, 34, 41, 75, 83, 134, 163–164, 179. *See also* Modernism, Naturalism; Pastoral; Realism; Romance; Sensationalism; Sentimentalism
Giant Power, 126–127
Gilman, Charlotte Perkins, 7–12, 14, 91–128, 180, 183, 212. *See also* Battery, as metaphor or analogy; Rest cure
fiction by, 113–127
nonfiction by, 8–9, 92–93, 97–109, 112–113
Gilmore, Paul, 3, 7
Glazener, Nancy, 7–8, 13, 134
Goble, Mark, 3, 7

Halliday, Sam, 3, 7, 226n17
Haraway, Donna, 94, 245n1
Heidegger, Martin, 231n13
Helmholtz, Hermann, 59, 72, 87, 111
Hilgartner, Stephen, 15, 218
History
composition of, 10–12, 16, 20, 44, 49, 57, 139, 167–168, 182, 192–196, 209
as usable, 16, 126, 167, 192, 205
and whitewashing, 181, 192, 197
Horkheimer, Max, 28, 235n14
Howells, William D., 14, 19, 45, 62–63, 68–69, 83, 226n18
and literary realism, 45, 83, 229n43

Howells, William D. (cont.)
    opinions regarding electric execution, 62–63
Hughes, Agatha C., 167
Hughes, Thomas P., 6, 126, 132, 167, 215
Humanism, 40, 168, 170–172, 181, 183–184, 187, 193, 195, 198–206, 215–217
Hybrids. *See* Genre, hybridizations of

Individualism or individuation, 8, 14–15, 46, 47, 88, 92, 95, 108, 144, 194
    as opposed to interconnection, 4–12, 15, 41, 58, 88–93, 98–102, 112–113, 120, 125–127, 150, 174, 196, 215 (*see also* Systems)
Interconnection. *See* Electricity, as interconnected; Individualism or individuation, as opposed to interconnection; Systems
Interdisciplinary methods, 5, 15, 168, 216
Interpretive flexibility, 129–134, 162, 206, 216

Jakobson, Roman, 246n4
Jasanoff, Sheila, 15, 217–218
Jefferson, Thomas, 142, 161
Johnson, Mark, 213–214
Jokes, 24–26, 30, 34, 226n18
Journalists, 56–58, 75, 78–87

Kemmler, William, 55, 67–74
Kim, Sang-Hyun, 15, 217–218
Kline, Ronald, 126

Labor, 38, 77, 103, 114, 119–121, 131–132, 143, 146–147, 149, 154–159, 161–162, 171, 174, 215
    as invisibilized, 146–147, 154, 176, 181, 215
    and the "labor-saving" device, 38, 121, 159, 176
    and social hierarchies, 77, 103, 119–121, 156, 158, 176
Lakoff, George, 213–214
Latour, Bruno, 149, 225n7, 232n19, 241n22
Layton, Edwin T., Jr., 131, 144–145
Lightning, 24–32, 60, 64–67, 93. *See also* Electricity, as natural
London, Jack, 7–8, 14–15, 127–167, 193
    biography of, 129–135
    Burning Daylight, 132–135, 140–155, 160–163
    Call of the Wild, 9, 15, 109, 128, 134, 145
    Martin Eden, 164–165
    The Valley of the Moon, 132–133, 155–163
Luhmann, Niklas, 8

Marvin, Carolyn, 226n18
Marx, Leo, 18, 20, 38, 44, 49, 88, 149, 169, 170
Maxwell, James Clerk, 180
Medical science. *See* Electricity, and medicine; Nerves
Metaphors, mixed, 9–12, 21–22, 26, 36, 48, 72–74, 94, 100, 104, 163. *See also* Genre, hybridizations of
Metonymy, 3, 6, 10, 13, 15, 22, 57, 88–89, 127, 156, 178, 213–214
Misa, Thomas J., 20, 175, 215–216
Modernism, literary, 12, 179, 212
Modernity and modernization, 3, 5, 10, 20, 22, 43–46, 73, 92, 101–102, 109, 138, 143, 157–158, 164, 170, 188–189, 191, 201, 209, 217
    as inherently interconnected, 5–7, 46, 101–102, 109, 112–113
Mumford, Lewis, 15, 160–161, 163–164, 167–185, 187, 192–198, 201–209, 214
    Golden Day, The, 168–171, 173–174, 192–197, 199
    Technics and Civilization, 168–171, 174, 178–183, 185

Index

Nadel, Alan, 192–193, 245n34
National Electric Light Association (NELA), 136–137, 151, 175, 240n12
Naturalism, literary, 15, 134, 143–146, 155–157, 160, 163–164
Nerves
  figurative depictions of, 4, 63, 84–85, 93–94, 99, 111–112, 141, 219
  scientific understanding of, 35, 59, 72, 93–94, 99, 102, 110–111, 219
Norris, Frank, 145
Norris, George W., 137
Nye, David E., 3, 7, 11, 51, 175, 215

Otis, Laura, 111

Paige Compositor, 17, 224n2
Pastoral, 5, 46, 132–134, 155, 157–159, 163
Pinchot, Gifford, 126–127
Pizer, Donald, 76, 78
Poe, Edgar Allan, 226n17
Posthumanism, 94, 101, 201
Progress
  conflation of social and electrical, 3, 10, 17, 19, 22, 38–39, 44–49, 60, 85, 94–96, 104–105, 107–109, 121, 139, 151–153, 164–165, 170–177, 188, 190, 205–206, 214
  and the construction of history, 10–12, 14, 49, 104–105, 139, 151–153
  and obsolescence, 30, 60, 107, 120

Quirk, John, 125

Race, 44–45, 102–104, 120–124, 140, 156, 159–160, 168, 172, 181, 184–185, 187, 198, 204. *See also* White supremacy
  and social power, 45, 102–103, 120–121, 124, 156, 159–160, 181, 184
Realism, literary, 16, 42, 45, 66, 75, 83, 134, 163–164, 226n14
Rest cure, 100–101, 122

Rockwell, A. D., 69–74, 110
Romance, 16, 25, 45, 64, 79, 83, 130–131, 163–164, 169, 192–197, 201, 209
  critiques of, 25, 45–46, 64, 79, 130–131, 140, 192, 195, 197
  and the emplotment of history, 181, 192–197
  Lewis Mumford's defense of, 163–164, 169–171, 193, 209, 219
  and the romantic individual, 46, 103, 108
Rural Electrification Administration, 162, 178, 186

Sarat, Austin, 51, 61, 68
Schatzberg, Eric, 18, 20, 147, 231n13
Scientific American, 59–62, 66–67, 72
Segal, Howard P., 124
Seltzer, Marc, 11
Sensationalism, 28, 34, 63, 67–68, 74–76, 79–81, 83
Sentimentality or sentimentalism, 64, 80, 83, 130, 133, 135
Slavery, in the American imagination, 37–40, 173–174, 176–183, 194, 197, 201, 203, 207–209, 214
Smith, Henry Nash, 19, 49
Smith, Merritt Roe, 170
Social construction of technology, 133–134, 162, 188, 217
Socialism, 8, 91, 95, 105, 144, 147, 163, 185
Sociotechnical imaginary, 15, 217–218
Spencer, Herbert, 134, 144–145
Steinmetz, Charles P., 14, 95–96, 104–106, 127–128
Sublime, 2, 21, 30, 34, 42, 64, 142
Systems, idea of, 4–12, 31, 39, 40, 46–47, 58, 84, 88–93, 98–102, 108–109, 121, 124, 126–127, 175
  as opposed to atomization, 4–12, 15, 41, 58, 88–93, 98–102, 112–113, 120, 125–127, 150, 174, 196, 215

Technological fallacy, 13, 20, 48–52, 56–57, 91, 133, 162, 167, 170, 214

Technology. *See also* Determinism; Interpretive flexibility; Social construction of technology
  as a concept or word, 13, 16, 18–20, 22, 24, 38, 48–52, 84, 91, 125, 149–150, 158, 162, 167, 170, 175, 213–214
  and humanism, 172, 183, 187, 201, 206

Telegraphy and telephony. *See* Communication systems

Tesla, Nikola, 17–18, 136, 144

Thomas de la Peña, Carolyn, 101–102

Tichi, Cecelia, 223n15

Toomer, Jean, 212–213

Transcendentalism, 5, 90, 197, 207

Twain, Mark, 1–3, 13, 17–50, 59–60, 73, 93–94, 212, 216–218, 226n18
  A Connecticut Yankee in King Arthur's Court, 17, 19, 21–50, 93–94
  "Battle of the Sand-belt," 35, 37, 41–45, 47, 59
  "Beginnings of Civilization," 21–23, 28, 32, 37, 44
  hybrid rhetorical technique, 20–22, 26, 30, 34, 36, 41–42, 48

Utilitarianism, 60, 164, 171, 194, 197, 201, 209, 215, 219

Utopia and utopianism
  in fiction, 39–40, 46, 68, 89, 91, 96, 113–128, 132, 196, 206, 208–209
  in nonfiction, 91, 94, 96, 113, 121, 124–127, 132, 152–153, 179, 208–209

Veblen, Thorstein, 84, 158, 171, 224n4, 231n13, 237n25

Venus electrificata experiment, 11, 191

Wajcman, Judy, 103, 236n18

Ward, Lester F., 14, 97, 105, 111–112

Waste, 60, 107, 121, 141, 237n25

Water power, 117, 119, 126, 137, 139, 145–146, 150, 152, 179–180

Wellman, Barry, 31

Wells, Ida B., 121

West, American, as inherently electrical, 141–142. *See also* Electrification, of the West

Westinghouse, George, 54–55, 104, 151

White, Hayden, 244n33

White supremacy, 102–103, 161, 184, 187, 197, 204

Whitman, Walt, 9–11, 98–99, 105, 113, 169, 183, 193, 208

Wilcox, Johnnie, 185

Williams, Raymond, 224n4

Williams, Rosalind, 16, 180, 228n35, 243n17, 246n6

Wimsatt, W. K., and Monroe Beardsley, 20

Women, 75, 81, 93, 101–103, 107, 113, 116–121, 156, 176, 217. *See also* Addams, Jane; Atherton, Gertrude; Consumers, as feminine; Gender; Gilman, Charlotte Perkins; Rest cure; Wells, Ida B.
  as actual or fictional experts, 103, 116–120, 204, 238n34
  and labor, 93, 103, 121, 156, 176, 236n18

Woolgar, Steve, 93, 149, 212, 241n22

World's fairs, 10, 90, 121, 138. *See also* Chicago Columbian Exposition

Wright, John S., 174, 196

INSIDE TECHNOLOGY

edited by Wiebe E. Bijker, W. Bernard Carlson, and Trevor Pinch

Jennifer L. Lieberman, *Power Lines: Electricity in American Life and Letters, 1882–1952*

Pablo J. Boczkowski and C. W. Anderson, editors, *Remaking the News: Essays on the Future of Journalism Scholarship in the Digital Age*

Benoît Godin, *Models of Innovation: History of an Idea*

Brice Laurent, *Democratic Experiments: Problematizing Nanotechnology and Democracy in Europe and the United States*

Stephen Hilgartner, *Reordering Life: Knowledge and Control in the Genomics Revolution*

Cyrus C. M. Mody, *The Long Arm of Moore's Law: Microelectronics and American Science*

Harry Collins, Robert Evans, and Christopher Higgins, *Bad Call: Technology's Attack on Referees and Umpires and How to Fix It*

Tiago Saraiva, *Fascist Pigs: Technoscientific Organisms and the History of Fascism*

Teun Zuiderent-Jerak, *Situated Intervention: Sociological Experiments in Health Care*

Basile Zimmermann, *Technology and Cultural Difference: Electronic Music Devices, Social Networking Sites, and Computer Encodings in Contemporary China*

Andrew J. Nelson, *The Sound of Innovation: Stanford and the Computer Music Revolution*

Sonja D. Schmid, *Producing Power: The Pre-Chernobyl History of the Soviet Nuclear Industry*

Casey O'Donnell, *Developer's Dilemma: The Secret World of Videogame Creators*

Christina Dunbar-Hester, *Low Power to the People: Pirates, Protest, and Politics in FM Radio Activism*

Eden Medina, Ivan da Costa Marques, and Christina Holmes, editors, *Beyond Imported Magic: Essays on Science, Technology, and Society in Latin America*

Anique Hommels, Jessica Mesman, and Wiebe E. Bijker, editors, *Vulnerability in Technological Cultures: New Directions in Research and Governance*

Amit Prasad, *Imperial Technoscience: Transnational Histories of MRI in the United States, Britain, and India*

Charis Thompson, *Good Science: The Ethical Choreography of Stem Cell Research*

Tarleton Gillespie, Pablo J. Boczkowski, and Kirsten A. Foot, editors, *Media Technologies: Essays on Communication, Materiality, and Society*

Catelijne Coopmans, Janet Vertesi, Michael Lynch, and Steve Woolgar, editors, *Representation in Scientific Practice Revisited*

Rebecca Slayton, *Arguments that Count: Physics, Computing, and Missile Defense, 1949–2012*

Stathis Arapostathis and Graeme Gooday, *Patently Contestable: Electrical Technologies and Inventor Identities on Trial in Britain*

Jens Lachmund, *Greening Berlin: The Co-Production of Science, Politics, and Urban Nature*

Chikako Takeshita, *The Global Biopolitics of the IUD: How Science Constructs Contraceptive Users and Women's Bodies*

Cyrus C. M. Mody, *Instrumental Community: Probe Microscopy and the Path to Nanotechnology*

Morana Alač, *Handling Digital Brains: A Laboratory Study of Multimodal Semiotic Interaction in the Age of Computers*

Gabrielle Hecht, editor, *Entangled Geographies: Empire and Technopolitics in the Global Cold War*

Michael E. Gorman, editor, *Trading Zones and Interactional Expertise: Creating New Kinds of Collaboration*

Matthias Gross, *Ignorance and Surprise: Science, Society, and Ecological Design*

Andrew Feenberg, *Between Reason and Experience: Essays in Technology and Modernity*

Wiebe E. Bijker, Roland Bal, and Ruud Hendricks, *The Paradox of Scientific Authority: The Role of Scientific Advice in Democracies*

Park Doing, *Velvet Revolution at the Synchrotron: Biology, Physics, and Change in Science*

Gabrielle Hecht, *The Radiance of France: Nuclear Power and National Identity after World War II*

Richard Rottenburg, *Far-Fetched Facts: A Parable of Development Aid*

Michel Callon, Pierre Lascoumes, and Yannick Barthe, *Acting in an Uncertain World: An Essay on Technical Democracy*

Ruth Oldenziel and Karin Zachmann, editors, *Cold War Kitchen: Americanization, Technology, and European Users*

Deborah G. Johnson and Jameson W. Wetmore, editors, *Technology and Society: Building Our Sociotechnical Future*

Trevor Pinch and Richard Swedberg, editors, *Living in a Material World: Economic Sociology Meets Science and Technology Studies*

Christopher R. Henke, *Cultivating Science, Harvesting Power: Science and Industrial Agriculture in California*

Helga Nowotny, *Insatiable Curiosity: Innovation in a Fragile Future*

Karin Bijsterveld, *Mechanical Sound: Technology, Culture, and Public Problems of Noise in the Twentieth Century*

Peter D. Norton, *Fighting Traffic: The Dawn of the Motor Age in the American City*

Joshua M. Greenberg, *From Betamax to Blockbuster: Video Stores and the Invention of Movies on Video*

Mikael Hård and Thomas J. Misa, editors, *Urban Machinery: Inside Modern European Cities*

Christine Hine, *Systematics as Cyberscience: Computers, Change, and Continuity in Science*

Wesley Shrum, Joel Genuth, and Ivan Chompalov, *Structures of Scientific Collaboration*

Shobita Parthasarathy, *Building Genetic Medicine: Breast Cancer, Technology, and the Comparative Politics of Health Care*

Kristen Haring, *Ham Radio's Technical Culture*

Atsushi Akera, *Calculating a Natural World: Scientists, Engineers and Computers during the Rise of US Cold War Research*

Donald MacKenzie, *An Engine, Not a Camera: How Financial Models Shape Markets*

Geoffrey C. Bowker, *Memory Practices in the Sciences*

Christophe Lécuyer, *Making Silicon Valley: Innovation and the Growth of High Tech, 1930–1970*

Anique Hommels, *Unbuilding Cities: Obduracy in Urban Sociotechnical Change*

David Kaiser, editor, *Pedagogy and the Practice of Science: Historical and Contemporary Perspectives*

Charis Thompson, *Making Parents: The Ontological Choreography of Reproductive Technology*

Pablo J. Boczkowski, *Digitizing the News: Innovation in Online Newspapers*

Dominique Vinck, editor, *Everyday Engineering: An Ethnography of Design and Innovation*

Nelly Oudshoorn and Trevor Pinch, editors, *How Users Matter: The Co-Construction of Users and Technology*

Peter Keating and Alberto Cambrosio, *Biomedical Platforms: Realigning the Normal and the Pathological in Late-Twentieth-Century Medicine*

Paul Rosen, *Framing Production: Technology, Culture, and Change in the British Bicycle Industry*

Maggie Mort, *Building the Trident Network: A Study of the Enrollment of People, Knowledge, and Machines*

Donald MacKenzie, *Mechanizing Proof: Computing, Risk, and Trust*

Geoffrey C. Bowker and Susan Leigh Star, *Sorting Things Out: Classification and Its Consequences*

Charles Bazerman, *The Languages of Edison's Light*

Janet Abbate, *Inventing the Internet*

Herbert Gottweis, *Governing Molecules: The Discursive Politics of Genetic Engineering in Europe and the United States*

Kathryn Henderson, *On Line and On Paper: Visual Representation, Visual Culture, and Computer Graphics in Design Engineering*

Susanne K. Schmidt and Raymund Werle, *Coordinating Technology: Studies in the International Standardization of Telecommunications*

Marc Berg, *Rationalizing Medical Work: Decision-Support Techniques and Medical Practices*

Eda Kranakis, *Constructing a Bridge: An Exploration of Engineering Culture, Design, and Research in Nineteenth-Century France and America*

Paul N. Edwards, *The Closed World: Computers and the Politics of Discourse in Cold War America*

Donald MacKenzie, *Knowing Machines: Essays on Technical Change*

Wiebe E. Bijker, *Of Bicycles, Bakelites, and Bulbs: Toward a Theory of Sociotechnical Change*

Louis L. Bucciarelli, *Designing Engineers*

Geoffrey C. Bowker, *Science on the Run: Information Management and Industrial Geophysics at Schlumberger, 1920–1940*

Wiebe E. Bijker and John Law, editors, *Shaping Technology / Building Society: Studies in Sociotechnical Change*

Stuart Blume, *Insight and Industry: On the Dynamics of Technological Change in Medicine*

Donald MacKenzie, *Inventing Accuracy: A Historical Sociology of Nuclear Missile Guidance*

Pamela E. Mack, *Viewing the Earth: The Social Construction of the Landsat Satellite System*

H. M. Collins, *Artificial Experts: Social Knowledge and Intelligent Machines*

http://mitpress.mit.edu/books/series/inside-technology